数学类专业英语简明教程

戴建新　郭小亚　朱　赟　赵君喜　编著

东南大学出版社
SOUTHEAST UNIVERSITY PRESS
·南京·

图书在版编目(CIP)数据

数学类专业英语简明教程 / 戴建新等编著. —南京：
东南大学出版社，2021.11
ISBN 978-7-5641-9787-2

Ⅰ. ①数… Ⅱ. ①戴… Ⅲ. ①数学-英语-高考学校
-教材 Ⅳ. ①O1

中国版本图书馆 CIP 数据核字(2021)第 231328 号

责任编辑:夏莉莉　　责任校对:张万莹　　封面设计:顾晓阳　　责任印制:周荣虎

数学类专业英语简明教程

编　著	戴建新　郭小亚　朱　赟　赵君喜	
出版发行	东南大学出版社	
社　址	南京市四牌楼 2 号(邮编:210096　电话:025-83793330)	
经　销	全国各地新华书店	
印　刷	广东虎彩云印刷有限公司	
开　本	700mm×1000mm　1/16	
印　张	7.5	
字　数	152 千字	
版　次	2021 年 11 月第 1 版	
印　次	2021 年 11 月第 1 次印刷	
书　号	ISBN　978-7-5641-9787-2	
定　价	28.00 元	

本社图书若有印装质量问题,请直接与营销部联系,电话:025-83791830。

前　　言

　　随着国家改革开放和科技进步,国际交流合作越来越频繁,很多高校将专业英语列入专业的人才培养方案,专业英语既是基础英语教学的延伸,又是基础英语与专业教学的结合与实践。同样,数学类专业英语课程是在国际交流广泛化和多样化的情况下,为了满足新形势的发展和一流本科专业的需要为数学类等专业的本科学生开设的一门专业必修课程,是学好数学专业和培养学生综合能力的一门重要工具课程。开设数学类专业英语课程,就是要突出语言的专业性、实用性和交际性特点,使学生不仅掌握基础英语语言知识,同时了解数学等专业性较强的学科领域内容。遵照数学类专业英语学习的科学规律特点,为培养多方面的人才,数学专业英语在内容上注重广泛性,尽量包含多个学科与分支,同时结合数学学科特点,有较多的基本概念、定义、定理和推理的表达和证明等内容,以便于打牢基础。各部分既要构成一个统一整体,同时又应自成体系。

　　本教材关注前沿,保留大量经典数学学科基本知识,适当增加了专业学科前沿知识,有利于跟踪国际学术发展的最新动态。本教材的目的是使学生掌握数学类专业英语专业词汇,熟悉主要的数学类专业英语文章类型及相关数学类专业知识,同时培养学生的阅读习惯和阅读技巧,提高学生的英汉翻译能力,以及让学生具备高级技术应用型人才所必需的数学类专业英语写作基础知识。

　　数学类专业英语教学强调通过自身的体验来构建对数学专业理论的认知,教材从某种程度上影响着这一教学模式的形成。数学专业英语教材编写充分体现“在运用中学习语言”的宗旨。本教材适用对象为数学类专业大学三、四年级的学生,还可供本科学生以及研究生在实际学习中参考阅读。使用本书作为教材的教师不必拘泥于本书编写的前后顺序,对于所讲授的内容也可根据课时和学生的实际情况有所侧重和取舍。同时,作为一门专业语言课程,教学必须与数学专业知识的介绍同步进行,有机结合。

Contents

LESSON ONE

TEXT A

The Dimension of a Vector Space

The next result, with its classical elegant proof, says that if a vector space V has a *finite* spanning set S, then the size of any linearly independent set cannot exceed the size of S.

Theorem 1. *Let V be a vector space and assume that the vectors v_1, $\cdots$, v_n are linearly independent and the vectors s_1, $\cdots$, s_m span V. Then $n \leqslant m$.*

Proof. First, we list the two sets of vectors: the spanning set followed by the linearly independent set

$$s_1, \cdots, s_m; v_1, \cdots, v_n$$

Then we move the first vector v_1 to the front of the first list

$$v_1, s_1, \cdots, s_m; v_2, \cdots, v_n$$

Since s_1, $\cdots$, s_m span V, v_1 is a linear combination of the $s_i's$. This implies that we may remove one of the $s_i's$, which by reindexing if necessary can be s_1, from the first list and still have a spanning set

$$v_1, s_2, \cdots, s_m; v_2, \cdots, v_n$$

Note that the first set of vectors still spans V and the second set is still linearly independent.

Now we repeat the process, moving v_2 from the second list to the first list v_1, v_2, s_2, $\cdots$, s_m; v_3, $\cdots$, v_n. As before, the vectors in the first list are linearly dependent, since they spanned V before the inclusion of v_2. However, since the $v_i's$ are linearly independent, any nontrivial linear combination of the vectors in the first list that equals 0 must involve at least one of the $s_i's$. Hence, we may remove that vector, which again by reindexing if necessary may be taken to be s_2 still have a spanning set

$$v_1, v_2, s_3, \cdots, s_m; v_3, \cdots, v_n$$

Once again, the first set of vectors spans V and the second set is still linearly independent.

Now, if $m < n$, then this process will eventually exhaust the $s'_i s$ and lead to the list

$$v_1, v_2, \cdots, v_m; v_{m+1}, \cdots, v_n$$

where $v_1, v_2, \cdots, v_m$ span V, which is clearly not possible since v_n is not in the span of $v_1, v_2, \cdots, v_m$. Hence, $n \leqslant m$.

Corollary 1. *If V has a finite spanning set, then any two bases of V have the same size.*

Now let us prove Corollary 1 for arbitrary vector spaces.

Theorem 2. *If V is a vector space, then any two bases for V have the same cardinality.*

Proof. We may assume that all bases for V are infinite sets, for if any basis is infinite then V has a finite spanning set and so Corollary 1 applies.

Let $\mathcal{B} = \{b_i \mid i \in I\}$ be a basis for V, indexed by the set I, and let $\mathcal{C}$ be another basis for V. Then any vector $c \in \mathcal{C}$ can be written as a finite linear combination of the vectors in $\mathcal{B}$, where all of the coefficients are nonzero, say

$$c = \sum_{i \in U_c} r_i b_i$$

But because $\mathcal{C}$ is a basis, we must have

$$\bigcup_{c \in \mathcal{C}} U_c = I$$

for if the vectors in $\mathcal{C}$ can be expressed as finite linear combinations of the vectors in a proper subset $\mathcal{B}'$ of $\mathcal{B}$, then $\mathcal{B}'$ spans V, which is not the case.

Since $|U_c| < \aleph_0$ for all $c \in \mathcal{C}$, A related theorem implies that

$$|\mathcal{B}| = |I| \leqslant \aleph_0 |\mathcal{C}| = |\mathcal{C}|$$

But we may also reverse the roles of $\mathcal{B}$ and $\mathcal{C}$, to conclude that $|\mathcal{B}| \leqslant |\mathcal{C}|$ and so $|\mathcal{B}| = |\mathcal{C}|$ by the Schröder-Bernstein theorem.

Theorem 2 allows us to make the following definition.

Definition. *A vector space V is **finite-dimensional** if it is the zero space $\{0\}$, or if it has a finite basis. All other vector spaces are **infinite-dimensional**. The **dimension** of the zero space is 0 and the **dimension** of any nonzero vector space V is the cardinality of any basis for V. If a vector space V has a basis of cardinality κ, we say that V is κ-**dimensional** and write $dim(V) = \kappa$.*

It is easy to see that if S is a subspace of V, then $\dim(S) \leqslant \dim(V)$. If in addition, $\dim(S) = \dim(V) < \infty$, then $S = V$.

Theorem 3. *Let V be a vector space.*

(1) *If $\mathcal{B}$ is a basis for V and if $\mathcal{B} = \mathcal{B}_1 \cup \mathcal{B}_2$ and $\mathcal{B}_1 \cap \mathcal{B}_2 = \varnothing$, then*

$$V = \langle \mathcal{B}_1 \rangle \oplus \langle \mathcal{B}_2 \rangle$$

(2) *Let* $V = S \oplus T$. *If* $\mathcal{B}_1$ *is a basis for* S *and* $\mathcal{B}_2$ *is a basis for* T, *then* $\mathcal{B}_1 \cap \mathcal{B}_2 = \varnothing$ *and* $\mathcal{B} = \mathcal{B}_1 \cup \mathcal{B}_2$ *is a basis for* V.

Theorem 4. *Let* S *and* T *be subspaces of a vector space* V. *Then*
$$\dim(S) + \dim(T) = \dim(S + T) + \dim(S \cap T)$$
In particular, if T *is any complement of* S *in* V, *then*
$$\dim(S) + \dim(T) = \dim(V)$$
that is,
$$\dim(S \oplus T) = \dim(S) + \dim(T)$$

Proof. Suppose that $\mathcal{B} = \{b_i \mid i \in I\}$ is a basis for $S \cap T$. Extend this to a basis $\mathcal{A} \cup \mathcal{B}$ for S where $\mathcal{A} = \{a_j \mid j \in J\}$, indexed by the set J, is disjoint from $\mathcal{B}$. Also, extend $\mathcal{B}$ to a basis $\mathcal{B} \cup \mathcal{C}$ for T where $\mathcal{C} = \{c_k \mid k \in K\}$, indexed by the set K, is disjoint from $\mathcal{B}$. We claim that $\mathcal{A} \cup \mathcal{B} \cup \mathcal{C}$ is a basis for $S + T$. It is clear that $\langle \mathcal{A} \cup \mathcal{B} \cup \mathcal{C} \rangle = S + T$.

To see that $\mathcal{A} \cup \mathcal{B} \cup \mathcal{C}$ is linearly independent, suppose to the contrary that
$$\alpha_1 v_1 + \cdots + \alpha_n v_n = 0$$
where $v_i \in \mathcal{A} \cup \mathcal{B} \cup \mathcal{C}$ and $\alpha_i \neq 0$ for all i. There must be vectors v_i in this expression from both $\mathcal{A}$ and $\mathcal{C}$, since $\mathcal{A} \cup \mathcal{B}$ and $\mathcal{B} \cup \mathcal{C}$ are linearly independent. Isolating the terms involving the vectors from $\mathcal{A}$ on one side of the equality shows that there is a nonzero vector in $x \in \langle \mathcal{A} \rangle \cap \langle \mathcal{B} \cup \mathcal{C} \rangle$. But then $x \in S \cap T$ and so $x \in \langle \mathcal{A} \rangle \cap \langle \mathcal{B} \rangle$, which implies that $x = 0$, a contradiction. Hence, $\mathcal{A} \cup \mathcal{B} \cup \mathcal{C}$ is linearly independent and a basis for $S + T$. Now,
$$\begin{aligned}
\dim(S) + \dim(T) &= |\mathcal{A} \cup \mathcal{B}| + |\mathcal{B} \cup \mathcal{C}| \\
&= |\mathcal{A}| + |\mathcal{B}| + |\mathcal{B}| + |\mathcal{C}| \\
&= |\mathcal{A}| + |\mathcal{B}| + |\mathcal{C}| + \dim(S \cap T) \\
&= \dim(S + T) + \dim(S \cap T)
\end{aligned}$$
as desired. It is worth emphasizing that while the equation
$$\dim(S) + \dim(T) = \dim(S + T) + \dim(S \cap T)$$
holds for all vector spaces, we cannot write
$$\dim(S + T) = \dim(S) + \dim(T) - \dim(S \cap T)$$
unless $S + T$ is finite-dimensional.

TEXT B

Singular Value Decomposition

The SVD is a very important decomposition which is used for many purposes other than solving least squares problems.

Theorem 1. *SVD. Let A be an arbitrary m-by-n matrix with $m \geqslant n$. Then we can write $A = U \Sigma V^T$, where U is m-by-n and satisfies $U^T U = I$, V is n-by-n and satisfies $V^T V = I$, and $\Sigma = \mathrm{diag}(\sigma_1, \cdots, \sigma_n)$, where $\sigma_1 \geqslant \cdots \geqslant \sigma_n \geqslant 0$.*

The columns $u_1, \cdots, u_n$ of U are called left singular vectors. The columns $v_1, \cdots, v_n$ of V are called right singular vectors. The σ_i are called singular values. (If $m < n$, the SVD is defined by considering A^T.)

A geometric restatement of this theorem is as follows. Given any m-by-n matrix A, think of it as mapping a vector $x \in \mathbb{R}^n$ to a vector $y = Ax \in \mathbb{R}^m$. Then we can choose one orthogonal coordinate system for $\mathbb{R}^n$ (where the unit axes are the columns of V) and another orthogonal coordinate system for $\mathbb{R}^m$ (where the unit axes are the columns of U) such that A is diagonal (Σ), i. e. maps a vector $x = \sum_{i=1}^{n} \beta_i V_i$ to $y = Ax = \sum_{i=1}^{n} \sigma_i \beta_i U_i$. In other words, any matrix is diagonal, provided we pick appropriate orthogonal coordinate systems for its domain and range.

Proof. We use induction on m and n: we assume that the SVD exists for $(m-1)$-by-$(n-1)$ matrices and prove it for m-by-n. We assume $A = 0$; otherwise we can take $\Sigma = 0$ and let U and V be arbitrary orthogonal matrices.

The basic step occurs when $n = 1$ (since $m \geqslant n$). We write $A = U \Sigma V^T$ with $U = A/\|A\|_2$, $\Sigma = \|A\|_2$, and $V = 1$.

For the induction step, choose v so $\|v\|_2 = 1$ and $\|A\|_2 = \|Av\|_2 > 0$. Such a v exists by the definition of $\|A\|_2 = \max_{\|v\|_2 = 1} \|Av\|_2$. Let $u = \frac{Av}{\|Av\|_2}$, which is a unit vector. Choose $\widetilde{U}$ and $\widetilde{V}$ to that $U = [u, \widetilde{U}]$ is an m-by-n orthogonal matrix, and $V = [v, \widetilde{V}]$ is an n-by-n orthogonal matrix. Now write

$$U^T AV = \begin{bmatrix} u^T \\ \widetilde{U}^T \end{bmatrix} \cdot A \cdot [v \quad \widetilde{V}] = \begin{bmatrix} u^T Av & u^T A \widetilde{V} \\ \widetilde{U}^T Av & \widetilde{U}^T A \widetilde{V} \end{bmatrix}.$$

Then

$$u^T A v = \frac{(Av)^T (Av)}{\|Av\|_2} = \frac{\|Av\|_2^2}{\|Av\|_2} = \|Av\|_2 = \|A\|_2 \equiv \sigma$$

and $\tilde{U}^T A v = \tilde{U}^T u \|Av\|_2 = 0$. We claim $u^T A \tilde{V} = 0$ too because otherwise $\sigma = \|A\|_2 = \|U^T A v\|_2 \geqslant \|[1,0,\cdots,0]U^T A V\|_2 = \|[\sigma \mid u^T A \tilde{V}]\|_2 > \sigma$.

So $U^T A V = \begin{bmatrix} \sigma & 0 \\ 0 & \tilde{U}^T A \tilde{V} \end{bmatrix} = \begin{bmatrix} \sigma & 0 \\ 0 & \tilde{A} \end{bmatrix}$. We may now apply the induction hypothesis to $\tilde{A}$ to get $\tilde{A} = U_1 \Sigma_1 V_1^T$, where U_1 is $(m-1)$-by-$(n-1)$, Σ_1 is $(n-1)$-by-$(n-1)$, and V_1 is $(n-1)$-by-$(n-1)$. So

$$U^T A V = \begin{bmatrix} \sigma & 0 \\ 0 & U_1 \Sigma_1 A_1^T \end{bmatrix} = \begin{bmatrix} 1 & 0 \\ 0 & U_1 \end{bmatrix} \begin{bmatrix} \sigma & 0 \\ 0 & \Sigma_1 \end{bmatrix} \begin{bmatrix} 1 & 0 \\ 0 & V_1 \end{bmatrix}^T$$

or

$$A = \left(U \begin{bmatrix} 1 & 0 \\ 0 & U_1 \end{bmatrix} \right) \begin{bmatrix} \sigma & 0 \\ 0 & \Sigma_1 \end{bmatrix} \left(V \begin{bmatrix} 1 & 0 \\ 0 & V_1 \end{bmatrix} \right)^T,$$

which is our desired decomposition.

The SVD has a large number of important algebraic and geometric properties, the most important of which we state here.

Theorem 2. *Let* $A = U \Sigma V^T$ *be the SVD of the m-by-n matrix* A, *where* $m \geqslant n$. *(There are analogous results for* $m < n$.*)*

(1) *Suppose that* A *is symmetric, with eigenvalues* λ_i *and orthonormal eigenvectors* u_i. *In other words* $A = U \Lambda U^T$ *is an eigendecomposition of* A, *with* $\Lambda = \mathrm{diag}(\lambda_1, \cdots, \lambda_n)$, $U = [u_1, \cdots, u_n]$, *and* $UU^T = I$. *Then an SVD of* A *is* $A = U \Sigma V^T$, *where* $\sigma_i = |\lambda_i|$ *and* $v_i = \mathrm{sign}(\lambda_i) u_i$, *where* $\mathrm{sign}(0) = 1$.

(2) *The eigenvalues of the symmetric matrix* $A^T A$ *are* σ_i^2. *The right singular vectors* v_i *are corresponding orthonormal eigenvectors.*

(3) *The eigenvalues of the symmetric matrix* AA^T *are* σ_i^2 *and m-n zeroes. The left singular vectors* u_i *are corresponding orthonormal eigenvectors for the eigenvalues* σ_i^2. *One can take any m-n other orthogonal vectors as eigenvectors for the eigenvalue 0.*

(4) *If* A *has full rank, the solution of* $\min_x \|Ax - b\|_2$ *is* $x = V \Sigma^{-1} U^T b$.

(5) $\|A\|_2 = \sigma_1$. *If* A *is also square and nonsingular, then* $\|A^{-1}\|_2^{-1} = \sigma_n$ *and*

$$\|A\|_2 \cdot \|A^{-1}\|_2 = \frac{\sigma_1}{\sigma_n}.$$

(6) *Suppose* $\sigma_1 \geqslant \cdots \geqslant \sigma_r \geqslant \sigma_{r+1} = \cdots = \sigma_n = 0$. *Then the rank of* A *is r. The null space of* A, *i.e. the subspace of vectors* v *such that* $Av = 0$ *is the space spanned by columns*

$r+1$ *through* n *of* V: span$(v_{r+1}, \cdots, v_n)$. *The range space of* A, *the subspace of vectors of the form* Aw *for all* w *is the space spanned by columns* 1 *through* r *of* U: span$(u_1, \cdots, u_r)$.

(7) *Write* $V = [v_1, v_2, \cdots, v_n]$ *and* $U = [u_1, u_2, \cdots, u_n]$, *so* $A = U \Sigma V^T = \sum_{i=1}^{n} \sigma_i u_i v_i^T$ (*a sum of rank-1 matrices*). *Then a matrix of rank* $k < n$ *closest to* A (*measured with* $\|\cdot\|_2$) *is* $A_k = \sum_{i=1}^{n} \sigma_i u_i v_i^T$ *and* $\|A - A_k\|_2 = \sigma_{k+1}$. *We may also write* $A_k = U \Sigma_k V^T$, *where* $\Sigma_k = \mathrm{diag}(\sigma_1, \cdots, \sigma_k, 0, \cdots, 0)$.

Proof.

(1) This is true by the definition of the SVD.

(2) $A^T A = V \Sigma U^T U \Sigma V^T = V \Sigma^2 V^T$. This is an eigendecomposition of $A^T A$, where the columns of V are the eigenvectors and the diagonal entries of Σ^2 are the eigenvalues.

(3) Choose an m-by-$(m-n)$ matrix $\tilde{U}$ so that $[U, \tilde{U}]$ is square and orthogonal. Then write

$$AA^T = U \Sigma V^T V \Sigma U^T = U \Sigma^2 U^T = [U, \tilde{U}] \begin{bmatrix} \Sigma^2 & 0 \\ 0 & 0 \end{bmatrix} [U, \tilde{U}]^T$$

This is an eigendecomposition of AA^T.

(4) $\|Ax - b\|_2^2 = \|U \Sigma V^T x - b\|_2^2$. Since A has full rank, so does Σ, and thus Σ is invertible. Now let $[U, \tilde{U}]^T$ be square and orthogonal as above so

$$\|U \Sigma V^T x - b\|_2^2 = \left\| \begin{bmatrix} U^T \\ \tilde{U}^T \end{bmatrix} (U \Sigma V^T x - b) \right\|_2^2$$

$$= \left\| \begin{bmatrix} \Sigma V^T x - U^T b \\ -\tilde{U}^T b \end{bmatrix} \right\|_2^2$$

$$= \| \Sigma V^T x - U^T b \|_2^2 + \| \tilde{U}^T b \|_2^2.$$

This is minimized by making the first term zero, i.e. $x = V \Sigma^{-1} U^T b$.

(5) It is clear from its definition that the two-norm of a diagonal matrix is the largest absolute entry on its diagonal.

$$\|A\|_2 = \|U^T AV\|_2 = \|\Sigma\|_2 = \sigma_1 \text{ and } \|A^{-1}\|_2 = \|V^T A^{-1} U\|_2 = \|\Sigma^{-1}\| = \sigma_n^{-1}.$$

(6) Again choose an m-by-n matrix $\tilde{U}$ so that the m-by-m matrix $\hat{U} = [U, \tilde{U}]$ is orthogonal. Since U and V are nonsingular, A and $\hat{U}^T AV = \begin{smallmatrix} n \times n \\ (m-n) \times n \end{smallmatrix} \equiv \hat{\Sigma}$ have the same rank-namely, r-by our assumption about Σ. Also, v is in the null space of A if and only if $V^T v$ is in the null space of $\hat{U}^T AV = \hat{\Sigma}$, since $Av = 0$ if and only if $\hat{U}^T AV(V^T v) = 0$ but

the null space of $\hat{\Sigma}$ is clearly spanned by columns $r + 1$ through n of the n-by-n identity matrix I_n, so the null space of A is spanned by V times these columns, i.e. v_{r+1} through v_n. A similar argument shows that the range space of A is the same as $\hat{U}$ times the range space of $\hat{U}^T A V = \hat{\Sigma}$, i. e. $\hat{U}$ times the first r columns of I_m, or u_1 through u_r.

(7) A_k has rank k by construction and

$$\|A - A_k\|_2 = \left\| \sum_{i=k+1}^{n} \sigma_i u_i v_i^T \right\| = \left\| U \begin{bmatrix} 0 & & & \\ & \sigma_{k+1} & & \\ & & \ddots & \\ & & & \sigma_n \end{bmatrix} V^T \right\| = \sigma_{k+1}.$$

It remains to show that there is no closer rank k matrix to A. Let B be any rank k matrix, so its null space has dimension $n - k$. The space spanned by $\{v_1, \cdots, v_{k+1}\}$ has dimension $k + 1$. Since the sum of their dimensions is $(n - k) + (k + 1) > n$, these two spaces must overlap. Let h be a unit vector in their intersection. Then

$$\|A - B\|_2^2 \geqslant \|(A - B)h\|_2^2 = \|Ah\|_2^2 = \|U \Sigma V^T h\|_2^2$$
$$= \|\Sigma (V^T h)\|_2^2$$
$$\geqslant \sigma_{k+1}^2 \|V^T h\|_2^2$$
$$= \sigma_{k+1}^2.$$

SENTENCE STRUCTURE ANALYSIS

1. The next result, with its classical elegant proof, says that if a vector space V has a finite spanning set S, then the size of any linearly independent set cannot exceed the size of S.

下一项结果的论证经典,它表明如果一个向量空间 V 包含一个有限的生成集 S,那么任何线性无关集的大小都不能超过 S 的大小。

这是一个宾语从句。that 引导的宾语从句作谓语动词 says 的宾语,其中"if a vector space V has a finite spanning set S"是条件状语从句,"then the size of any linearly independent set cannot exceed the size of S"表示在前句所述的条件下后句内容得以实现或是前句得到延续。

If..., then... 译为"若……则……""如果……那么……",这是数学文章中常见句型。

Eg.

1) If $\alpha > -1$, then the integral exists.

若 $\alpha > -1$,则积分存在。

2）If f and g are measurable functions, then so are $f+g$ and $f \cdot g$.

如果 f 和 g 是可测函数，那么 $f+g$ 和 $f \cdot g$ 也是可测函数。

2. The SVD is a very important decomposition which is used for many purposes other than solving least squares problems.

奇异值分解是一种非常重要的分解方法，除了求解最小二乘问题外，它还有很多用途。

这句话的中心句是 The SVD is a very important decomposition。定语从句 which is used for many purposes other than solving least squares problems 修饰先行词 decomposition。其中短语 other than 译为"除了……"。

Eg. Each member of G other than g is…

除 g 外 G 中的每个元素是……

3. It is clear from its definition that the two-norm of a diagonal matrix is the largest absolute entry on its diagonal.

从它的定义可以清楚地看出，对角矩阵的二范数是其对角线上绝对值最大的项。

It is clear that… 是主语从句最常见的一种结构。该句型为了防止句子头重脚轻，通常把形式主语 it 放在句首位置，that 引导的从句作为真正的主语置于句末，常译为"……是清楚的"。

Eg. It is not clear whether the result is sharp.

不清楚这个结果是不是最优的。

类似的句式还有：

It is evident/obvious that…　　　　　……是明显的/显而易见的。

It is easily seen that…　　　　　……是容易得到的。

It is easy to check that…　　　　　……是容易检验的。

WORDS AND PHRASES

TEXT A

dimension 维数	vector space 向量空间
linearly independent 线性无关的	set 集，集合
span 张成，生成	linear combination 线性组合
nontrivial 非平凡的	finite 有限的
base 基；基数	arbitrary 任意的

corollary 推论

theorem 定理

cardinality 基数；势

assume 假设

basis 基

coefficient 系数

nonzero 非零的

proper subset 真子集

finite-dimensional 有限维的

subspace 子空间

complement 补集

disjoint 不相交的；互斥的

contrary 对立的，相反的；相反命题

expression 表达式

contradiction 矛盾

hold for 适用于

TEXT B

singular value 奇异值

decomposition 分解

least squares problem 最小二乘问题

column 列

geometric 几何的

mapping 映射

orthogonal 正交的

coordinate system 坐标系

axes(axis 的复数) 轴

diagonal 对角的

matrix 矩阵

domain 定义域

range 值域

induction 归纳法

induction hypothesis 归纳假设

algebraic 代数的

analogous 类似的；类比的

symmetric 对称的

eigenvalue 特征值

orthonormal 正交的；正规化的

eigenvector 特征向量

eigendecomposition 特征分解

full rank 满秩

solution 解

square 方形的；正方形的

non-singular 非奇异的

null space 零空间

unit sphere 单位球面

image 像

ellipsoid 椭圆体

origin 原点

minimized 最小的

two-norm 二范数

diagonal matrix 对角矩阵

absolute 绝对的

entry （矩阵的）项

multiply 乘

principal axes 主轴

identity matrix 单位矩阵

intersection 交集

LESSON TWO

TEXT A

Limits and Convergence of Functions

So far we have considered the limits of sequences of real numbers. These sequences are real-valued functions defined on $\mathbb{Z}^+$ or $\mathbb{N}$. We now consider real-valued functions defined on a non-empty subset A of $\mathbb{R}$. It is useful to make definitions for a general set A, but the reader should have in mind examples such as an open interval, a closed interval, the set $\mathbb{Q}$ of rational numbers and the set $\{1/n : n \in \mathbb{N}\}$.

The notion of limit extends naturally to this setting. Suppose that $f: A \to \mathbb{R}$ is a function, that b is a limit point of A (which may or may not be an element of A) and that $l \in \mathbb{R}$. We say that $f(x)$ *converges to* l, or *tends to* l, as x to b if whenever $\epsilon > 0$ there exists $\delta > 0$ (which usually depends on ϵ) such that $|f(x) - l| < \epsilon$ for those $x \in A$ for which $0 < |x - b| < \delta$ (that is, for $x \in N_\delta^*(b) \cap A$). That is to say, as x gets close to b, $f(x)$ gets close to l. We say that l is the *limit of f as x tends to* b, write "$f(x) \to l$ as $x \to b$" and also write $l = \lim\limits_{x \to b} f(x)$. Note that in the case where $b \in A$, we do not consider the value of $f(b)$, but only the values of f at points nearby.

We now have the following elementary results. We say that f is *bounded* on A if the image set $f(A)$ is a bounded set.

Theorem 1. *Suppose that f, g and h are real-valued functions on a subset A of $\mathbb{R}$ and that b is a limit point of A.*

(1) *If $f(x) \to l$ as $x \to b$ and $f(x) \to m$ as $x \to b$, then $l = m$.*

(2) *If $f(x) \to l$ as $x \to b$ then there exists $\delta > 0$ such that f is bounded on $N_\delta^*(b) \cap A$.*

(3) *If $f(x) = l$ for all $x \in A$, then $f(x) \to l$ as $x \to b$.*

(4) *If $f(x) \to 0$ as $x \to b$, and $g(x)$ is bounded on $N_\delta^*(b) \cap A$ for some $\delta > 0$, then $f(x)g(x) \to 0$ as $x \to b$.*

(5) *If $f(x) \to l$ and $g(x) \to m$ as $x \to b$, then $f(x) + g(x) \to l + m$ as $x \to b$.*

(6) *If $f(x) \to l$ as $x \to b$ and $c \in \mathbb{R}$, then $cf(x) \to cl$ as $x \to b$.*

(7) *If* $f(x) \to l$ *and* $g(x) \to m$ *as* $x \to b$, *then* $f(x)g(x) \to lm$ *as* $x \to b$.

(8) *If* $f(x) \neq 0$ *for* $x \in A$, $l \neq 0$ *and* $f(x) \to l$ *as* $x \to b$, *then* $1/f(x) \to 1/l$ *as* $x \to b$.

(9) *If* $f(x) \to l$ *and* $g(x) \to m$ *as* $x \to b$, *and if* $f(x) \leqslant g(x)$ *for all* $x \in N_\delta^*(b) \cap A$ *for some* $\delta > 0$, *then* $l \leqslant m$.

(10) (*The sandwich principle*) *Suppose that* $f(x) \leqslant g(x) \leqslant h(x)$ *for all* $x \in N_\delta^*(b) \cap A$, *for some* $\delta > 0$, *and that* $f(x) \to l$ *and* $h(x) \to l$ *as* $x \to b$, *then* $g(x) \to l$ *as* $x \to b$.

Proof. Since the definition of limit is so similar to that of the limit of a sequence, the proofs are simple modifications of the proofs of corresponding results for sequences. The details are left as exercises for the readers.

Note that in several cases we have restricted attention to the behavior of f in a set $x \in N_\delta^*(b) \cap A$. This is clearly appropriate, since we are only concerned with the behavior of f as x approaches b.

It is a useful fact that we can characterize convergence in terms of convergent sequences.

Proposition 1. *Suppose that* f *is a real-valued function on a subset* A *of* $\mathbb{R}$, *that* b *is a limit point of* A *and that* $l \in \mathbb{R}$. *Then* $f(x) \to l$ *as* $x \to b$ *if and only if whenever* $(a_n)_{n=0}^\infty$ *is a sequence in* $A \setminus \{b\}$ *which tends to* b *as* $n \to \infty$, *then* $f(a_n) \to l$ *as* $n \to \infty$.

Proof. Suppose that $f(x) \to l$ as $x \to b$ and that $(a_n)_{n=0}^\infty$ is a sequence in $A \setminus \{b\}$ which tends to b as $n \to \infty$. Given $\epsilon > 0$, there exists $\delta > 0$ such that if $x \in N_\delta^*(b) \cap A$ then $|f(x) - l| < \epsilon$. There then exists n_0 such that $|a_n - b| < \delta$ for $n \geqslant n_0$. Then $|f(a_n) - l| < \epsilon$ for $n \geqslant n_0$, so that $f(a_n) \to l$ as $n \to \infty$.

Suppose that $f(x)$ does not converge to l as $x \to b$. Then there exists $\epsilon > 0$ for which we can find no suitable $\delta > 0$. If $n \in \mathbb{N}$, then $1/n$ is not suitable, and so there exists $x \in N_{1/n}^*(b) \cap A$ with $|f(x_n) - l| \geqslant \epsilon$. Then $x_n \to x$ as $n \to \infty$ and $f(x_n)$ does not converge to l as $n \to \infty$.

We have the following general principle of convergence.

Theorem 2. *Suppose that* f *is a real-valued function on a subset* A *of* $\mathbb{R}$, *that* b *is a limit point of* A *and that* $l \in \mathbb{R}$. *Then the following are equivalent.*

(1) *There exists* l *such that* $f(x) \to l$ *as* $x \to b$.

(2) *Whenever* $(a_n)_{n=0}^\infty$ *is a sequence in* $A \setminus \{b\}$ *which tends to* b *as* $n \to \infty$, *then* $(f(a_n))_{n=0}^\infty$ *is a Cauchy sequence.*

(3) *Given* $\epsilon > 0$ *there exists* $\delta > 0$ *such that if* x, $y \in N_\delta^*(b)$, *then* $|f(x) - f(y)| < \epsilon$.

Proof. Suppose that (1) holds, and that $(a_n)_{n=0}^{\infty}$ is a sequence in $A\setminus\{b\}$ which tends to b as $n\to\infty$. By Proposition 1, $f(a_n)\to l$ as $n\to\infty$. Since a convergent sequence is a Cauchy sequence, (2) holds.

Suppose that (3) fails. Then there exists $\epsilon>0$ for which for each $n\in\mathbb{N}$ there exists a_n, $a'_n\in N_{1/n}^*(b)\cap A$ with $|f(a_n)-f(a'_n)|\geqslant\epsilon$. Let $c_{2n-1}=a_n$ and $c_{2n}=a'_n$, for $n\in\mathbb{N}$. Then $c_n\to b$ as $n\to\infty$, and $(f(c_n))_{n=0}^{\infty}$ is not a Cauchy sequence. Thus (2) fails: (2) implies (3).

Finally suppose that (3) holds, and that $\epsilon>0$. There exists $\delta>0$ such that if x, $y\in N_\delta^*(b)\cap A$, then $|f(x)-f(y)|<\epsilon/2$. Suppose that $(a_n)_{n=0}^{\infty}$ is a sequence in $A\setminus\{b\}$ which tends to b as $n\to\infty$. Then there exists n_0 such that $a_n\in N_\epsilon^*(b)$ for $n\geqslant n_0$. Thus if m, $n\geqslant n_0$, then $|f(a_n)-f(a_m)|<\epsilon/2$, and so $(f(a_n))_{n=0}^{\infty}$ is a Cauchy sequence. By the general principle of convergence, there exists l such that $f(a_n)\to l$ as $n\to\infty$, and if $n\geqslant n_0$, then $|f(a_n)-l|\leqslant\epsilon/2$. Thus if $x\in N_\delta^*(b)\cap A$, then $|f(x)-l|\leqslant |f(x)-f(a_{n_0})|+|f(a_{n_0}-l)|<\epsilon$; $f(x)\to l$ as $x\to b$. Thus (3) implies (1).

We now turn to a result which corresponds to a related theorem. First we must introduce the idea of one-sided convergence. Suppose that f is a real-valued function on A and that $b\in\mathbb{R}$. Let $A_+=A\cap(b,\infty)$ and let $A_-=A\cap(-\infty,b)$. Suppose that b is a limit point of A_+, that is, $(b,b+\delta)$ is non-empty, for each $\delta>0$. Then we say that $f(x)$ *tends to l as $x\to b$ from the right* if whenever $\epsilon>0$ there exists $\delta>0$ such that if $x\in A\cap(b,b+\delta)$ then $|f(x)-l|<\epsilon$. We then write $f(x)\to l$ as $x\searrow b$, and denote l by $\lim_{x\searrow b}f(x)$, or, more briefly, by $f(b+)$. Similarly if $f(x)$ tends to l as $x\to b$ from the left, we denote the limit l by $\lim_{x\nearrow b}f(x)$, or $f(b-)$. Why do we use this terminology? If we consider the graph of f, drawn in the usual way, the variable x increases from left to right, and the values that the function f takes increase in an upwards direction. We therefore use "left" and "right" for the variable x, and reserve words such as "upper" or "lower" for the values of the function.

Theorem 3. *Suppose that f is a real-valued increasing function on A and that b is a limit point of $A_+=A\cap(b,\infty)$. If f is bounded below on A_+, then $f(x)\to\inf\{f(y): y\in A_+\}$ as $x\to b$ from the right.*

Similar results hold for "convergence from the left", and for decreasing functions.

Proof. Let $l=\inf\{f(y): y\in A_+\}$. Suppose that $\epsilon>0$. Then $l+\epsilon$ is not a lower bound for f on A_+, and so there exists $a\in A_+$ with $f(a)<1+\epsilon$. Let $\delta=a-b$. If $x\in A\cap(b,b+\delta)$ then $l\leqslant f(x)\leqslant f(a)<1+\epsilon$, so that $f(x)\to l$ as $x\to b$ from the right.

Corollary 1. *If b is a limit point of A_+ and A_-, then $f(b-) \leqslant f(b+)$.*

Proof. For $\sup\{f(x) : x \in A_-\} \leqslant \inf\{f(x) : x \in A^+\}$.

Suppose again that b is a limit point of a subset A of $\mathbb{R}$, and suppose that f is a real-valued function which is bounded on $N_\delta^*(b) \cap A$, for some $\delta < 0$. We can then define the upper and lower limits of f at b. For $0 < t < \delta$, let $M(t) = \sup\{f(x) : x \in N_t^*(b)\}$. Then $M(t)$ is an increasing function on $(0, \delta)$ which is bounded below. By Theorem 3 it follows that $M(t)$ converges to $M(0+) = \inf\{M(s) : 0 < s < \delta\}$ as $t \searrow 0$. $M(0+)$ is the *upper limit* or *limes superior* of f at b, and is denoted by $\limsup_{x \to b} f(x)$. The *lower limit*, or *limes inferior* $\liminf_{x \to b} f(x)$ is defined in a similar way.

Theorem 4. *Suppose that b is a limit point of a subset A of $\mathbb{R}$, and suppose that f is a real-valued function which is bounded on $N_\delta^*(b) \cap A$, for some $\delta > 0$. Then $f(x) \to l$ as $x \to b$ if and only if $\limsup_{x \to b} f(x) = \liminf_{x \to b} f(x) = l$.*

Proof. As an example, suppose that $x \in (0, 1]$. If $0 < 1/(n+1) < x \leqslant 1/n$, set $f(x) = n(n+1)(x - 1/(n+1))$. Then $\limsup_{x \to 0} f(x) = 1$ and $\liminf_{x \to 0} f(x) = 0$. The function f does not tend to a limit as $x \to 0$, but oscillates between the values 0 and 1.

We can also consider limits as $x \to +\infty$ or as $x \to -\infty$. Suppose that A is a subset of $\mathbb{R}$ which is not bounded above, that f is a real-valued function on A and that $l \in \mathbb{R}$. Then we say that $f(x) \to l$ as $x \to +\infty$ if whenever $\epsilon > 0$, there exists $x_0 \in \mathbb{R}$ such that if $x \in A$ and $x \geqslant x_0$, then $|f(x) - l| < \epsilon$. Similarly, if there exists x_0 such that f is bounded on $A \cap [x, \infty)$, and we define $M(x) = \sup\{f(a) : a \in A \cap [x, \infty)\}$ for $x \geqslant x_0$, then $M(x)$ is a decreasing function on $[x_0, \infty)$ which is bounded below; we define $\limsup_{x \to \infty} f(x)$ as $\inf\{M(x) : x \in [x_0, \infty)\}$. The lower limit is defined similarly. Limits as $x \to -\infty$ are defined in the same way. The reader should verify that all the results of this section, with appropriate modifications, extend these situations to without difficulty.

TEXT B

Riemann Integrable Functions

We say that a bounded function f on $[a, b]$ is *Riemann integrable* if its upper and lower integrals are equal. The common value is then the *Riemann integral* $\int_a^b f(x)\,dx$. In this expression, f is called the *integrand*. First, we must check that this extends the elementary integral of step functions.

Proposition 1. *If f is a step function, then it is Riemann integrable, and the*

Riemann integral is the same as the elementary integral.

Proof. Let E be the elementary integral. Since f is in both U_f and L_f,

$$E \leqslant \int_{\underline{a}}^{b} f(x) \, dx \leqslant \overline{\int_{a}^{b}} f(x) \, dx \leqslant E,$$

and so all the quantities are equal.

Proposition 2. *Suppose that f is a bounded function on $[a,b]$. Then f is Riemann integrable if and only if given $\epsilon > 0$, there exist step functions g and h with $h \leqslant f \leqslant g$ and $\int_{a}^{b} g(x) \, dx - \int_{a}^{b} h(x) \, dx < \epsilon$.*

Proof. This follows immediately from the definition.

Proposition 3. *Suppose that f is a bounded function on $[a,b]$. Then f is Riemann integrable if and only if given $\epsilon > 0$, there exists a dissection D such that $S_D - s_D < \epsilon$.*

Proof. The condition is clearly sufficient. If f is Riemann integrable and $\epsilon > 0$ then there exist dissections D_1 and D_2 such that $s_{D_1} + \epsilon/2 > \int_{a}^{b} f(x) \, dx > S_{D_2} - \epsilon/2$. Let $D = D_1 \bigvee D_2$. Then

$$S_D \leqslant S_{D_2} \leqslant s_{D_1} + \epsilon \leqslant s_D + \epsilon.$$

We can express this proposition in terms of the oscillation of f. Suppose that f is a bounded real-valued function on a non-empty set S. The *oscillation* $\Omega = \Omega(f, S)$ *of f on S* is defined as

$$\Omega(f, S) = \sup\{|f(s) - f(t)| : s, t \in S\} = \sup_{s \in S} f(s) - \inf_{s \in S} f(s).$$

Corollary 1. *Suppose that f is a bounded function on $[a,b]$. Then f is Riemann integrable if and only if given $\epsilon > 0$ there exists a dissection $D = \{a = x_0 < \cdots < x_k = b\}$ of $[a,b]$, with intervals $I_1, \cdots, I_k$ such that*

$$\sum_{j=1}^{k} \Omega(f, I_j)(x_j - x_{j-1}) < \epsilon.$$

Proof. For $S_D - s_D = \sum_{j=1}^{k} \Omega(f, I_j)(x_j - x_{j-1})$.

Corollary 2. *Suppose that f is a bounded function on $[a,b]$. Then f is Riemann integrable if and only if given $\epsilon > 0$ there exists a dissection $D = \{a = x_0 < \cdots < x_k = b\}$ of $[a,b]$ and a partition $G \cup B$ of $\{1, \cdots, k\}$ such that*

$$\Omega(f, I_j) \leqslant \epsilon \text{ for } j \in G \text{ and } \sum_{j \in B} l(I_j) < \epsilon.$$

where $I_1, \cdots, I_k$ are the intervals of the dissection.

Proof. Suppose that the condition is satisfied, and that $\epsilon > 0$. Let $\eta = \epsilon/(b - a +$

$\Omega(f,[a,b]))$. Then

$$\sum_{j=1}^{k} \Omega(f, I_j)(x_j - x_{j-1})$$

$$= \sum_{j \in G} \Omega(f, I_j)(x_j - x_{j-1}) + \sum_{j \in B} \Omega(f, I_j)(x_j - x_{j-1})$$

$$\leqslant \left(\sup_{j \in G} \Omega(f, I_j) \right) \sum_{j \in G} (x_j - x_{j-1}) + \Omega(f,[a,b]) \sum_{j \in B} (x_j - x_{j-1})$$

$$\leqslant (b-a)\eta + \Omega(f,[a,b])\eta = \epsilon,$$

so that f is Riemann integrable. Suppose conversely that f is Riemann integrable. By the previous corollary there exists a dissection D with

$$\sum_{j=1}^{k} \Omega(f, I_j)(x_j - x_{j-1}) < \min(\epsilon, \epsilon^2).$$

Let $G = \{j \in D : \Omega(f, I_j) \leqslant \epsilon\}$ and let $B = \{j \in D : \Omega(f, I_j) > \epsilon\}$. Then $\epsilon \sum_{j \in B} l(I_j) \leqslant$

$\sum_{j \in B} \Omega(f, I_j) l(I_j) < \epsilon^2$, which give the result. Many important functions are Riemann integrable.

Theorem 1. (1) *A continuous function on* $[a,b]$ *is Riemann integrable.*

(2) *A monotonic function on* $[a,b]$ *is Riemann integrable.*

Proof. In both cases, we use Proposition 3.

(1) We use the fact that f is uniformly continuous. Suppose that $\epsilon > 0$. There exists $\delta > 0$ such that if $|x - y| < \delta$, then $|f(x) - f(y)| < \epsilon/(b-a)$. Choose N so that $(b-a)/N < \delta$, and let D_N be the dissection of $[a,b]$ into N intervals $I_1, \cdots, I_N$ of equal length. Then $l(I_j) = (b-a)/N < \delta$, so that $M_j = \sup\{f(x) : x \in I_j\} < \inf\{f(x) : x \in I_j\} + \epsilon/(b-a) = m_j + \epsilon/(b-a)$, for $1 \leqslant j \leqslant N$, and so $S_{D_N} < s_{D_N} + \epsilon$.

(2) Without loss of generality we can suppose that f is increasing. Suppose that $\epsilon > 0$. Choose N so that $N > (f(b) - f(a))(b-a)/\epsilon$. Let D_N be the dissection of $[a, b]$ into N intervals of equal length, as before, and let $a = x_0 < x_1 < \cdots < x_N = b$ be the points of dissection. Then $m_j \geqslant f(x_{j-1})$ and $M_j = f(x_j)$, so that

$$S_D = \sum_{j=1}^{N} f(x_j) \frac{b-a}{N} = \sum_{j=1}^{N} f(x_{j-1}) \frac{b-a}{N} + (f(b) - f(a)) \frac{b-a}{N} < s_D + \epsilon.$$

As an easy example, let us calculate $\int_0^a x dx$, where $a > 0$. Then

$$S_{D_N} = \sum_{j=1}^{N} \left(\frac{aj}{N} \right) \left(\frac{a}{N} \right) = \frac{a^2}{N^2} \sum_{j=1}^{N} j = \frac{a^2}{2} \left(1 + \frac{1}{N} \right),$$

and

$$s_{D_N} = \sum_{j=1}^{N} \left(\frac{a(j-1)}{N}\right)\left(\frac{a}{N}\right) = \frac{a^2}{N^2}\sum_{j=1}^{N}(j-1) = \frac{a^2}{2}\left(1-\frac{1}{N}\right),$$

from which it follows that $\int_{0}^{a} x\,\mathrm{d}x = a^2/2$.

We can also characterize Riemann integrability in terms of a sequence of dissections.

Proposition 4. *Suppose that* $(D_r)_{r=1}^{\infty}$ *is a sequence of dissections of* $[a,b]$, *and that* $\delta(D_r)\to 0$ *as* $r\to\infty$. *If* f *is a bounded function on* $[a,b]$, *then* f *is Riemann integrable if and only if* $S_{D_r} - s_{D_r}\to 0$ *as* $r\to\infty$. *If so, then* $\int_{a}^{b} f(x)\,\mathrm{d}x = \lim_{r\to\infty} S_{Dr} = \lim_{r\to\infty} s_{Dr}$.

Corollary 3. *Suppose that* f *is a bounded function on* $[a,b]$ *and that* $J\in\mathbb{R}$. *Then the following are equivalent.*

(1) f *is Riemann integrable, and* $\int_{a}^{b} f(x)\,\mathrm{d}x = J$.

(2) *If* $(D_r)_{r=1}^{\infty}$ *is a sequence of dissections of* $[a,b]$ *with* $\delta(D_r)\to 0$ *as* $r\to\infty$, *if* $I_{r,1},\cdots,I_{r,q_r}$ *are the intervals of the dissection* D_r, *and if* $y_{r,p}\in I_{r,p}$, *for* $1\leqslant p\leqslant q_r$, *then*

$$\sum_{p=1}^{q_r} f(y_{r,p})l(I_{r,p})\to J \text{ as } r\to\infty.$$

It is important that (2) must hold for every choice of $y_{r,p}\in I_{r,p}$, and not just for one particular choice.

Proof. If f is Riemann integrable, and $\int_{a}^{b} f(x)\,\mathrm{d}x = J$, then

$$s_{D_r}\leqslant \sum_{p=1}^{q_r} f(y_{r,p})l(I_{r,p})\leqslant S_{Dr},$$

and so (2) follows from the proposition and the sandwich principle.

Conversely, suppose that (2) holds. For each $r\in\mathbb{N}$ and each p with $1\leqslant p\leqslant q_r$ there exist $y_{r,p}$ and $z_{r,p}$ in $I_{r,p}$ for which $f(y_{r,p}) - f(z_{r,p})\geqslant \Omega(f,I_{r,p})/2$. Then

$$0\leqslant \sum_{p=1}^{q_r} \Omega(f,I_{r,p})l(I_{r,p})\leqslant 2\sum_{p=1}^{q_r} (f(y_{r,p}) - f(z_{r,p}))l(I_{r,p}).$$

But

$$\sum_{p=1}^{q_r} (f(y_{r,p}) - f(z_{r,p}))l(I_{r,p})\to 0 \text{ as } r\to\infty,$$

and so

$$\sum_{p=1}^{q_r} \Omega(f,I_{r,p})l(I_{r,p})\to 0 \text{ as } r\to\infty.$$

Thus f is Riemann integrable, by Corollary 1. Further, since

$$s_{Dr} \leqslant \sum_{p=1}^{q_r} f(y_{r,p}) l(I_{r,p}) \leqslant S_{D_r},$$

it follows from the sandwich principle that

$$J = \lim_{r \to \infty} \left(\sum_{p=1}^{q_r} f(y_{r,p}) l(I_{r,p}) \right) = \int_a^b f(x) \, dx.$$

Let us consider some examples.

Example 1. A bounded function which is not Riemann integrable.

Let $f(x) = 1$ if x is rational, and $f(x) = 0$ if x is irrational. If $g = \sum_{j=0}^{k} v_j \chi_j \in U_f$,

then each I_j contains a rational number, and so $v_j \geqslant 1$. Thus $\int_a^b g(x) \, dx \geqslant b - a$, and so

$\overline{\int_a^b} f(x) \, dx \geqslant b - a$. Since $\chi_{[a,b]} \in U_f$, $\overline{\int_a^b} f(x) \, dx = b - a$. Similarly, if $h = \sum_{j=0}^{k} w_j \chi_j \in L_f$,

then each I_j contains an irrational number, and so $w_j \leqslant 0$. Thus $\int_a^b h(x) \, dx \leqslant 0$, and so

$\underline{\int_a^b} f(x) \, dx \leqslant 0$. Since $0 \in L_f$, $\underline{\int_a^b} f(x) \, dx = 0$. Thus f is not Riemann integrable.

Example 2. A Riemann integrable function on $[0,1]$ which is discontinuous at the rational points of $[0,1]$.

If $r \in [0,1]$ is rational, and $r = p/q$ in lowest terms, let $g(r) = 1/q$ and if x is irrational, let $g(x) = 0$. Then g is discontinuous at every rational number.

Suppose that $\epsilon > 0$. Then there exists q_0 such that $1/q_0 < \epsilon$. Then in a closed interval $[a,b]$, there are only finitely many rational numbers $r = p/q$ with $q \leqslant q_0$, so that $L = \{x \in [a,b]: g(x) > \epsilon\}$ is finite. We can include L in a finite set of intervals of total length less than ϵ: there exists a dissection

$$D = \{0 = x_0 < y_0 < x_1 < y_1 < \cdots < x_k < y_k = 1\}$$

such that

$$L \subseteq [x_0, y_0] \cup (x_1, y_1] \cup \cdots \cup [x_k, y_k],$$

and $\sum_{i=0}^{k} y_k - x_k < \epsilon$. If we take $G = \{x_i: 1 \leqslant i \leqslant k\}$ and $B = \{y_i: 0 \leqslant i \leqslant k\}$, then $\Omega(g,$

$I_j) \leqslant \epsilon$ for $j \in G$ and $\sum_{j \in B} l(I_j) < \epsilon$, so that g is Riemann integrable, by Corollary 2.

Further, $S_D < \epsilon(b-a) + \epsilon$, so that $\overline{\int_a^b} g(x) \, dx \leqslant 0$. Since g is non-negative, $\int_a^b g(x) \, dx = 0$.

Example 3. A function which is constant on a dense open subset of $[0,1]$, but which is not Riemann integrable.

Let $C^{(\epsilon)}$ be a fat Cantor set. $C^{(\epsilon)}$ is a perfect subset of $[0,1]$ with empty interior. Let $I_C(\epsilon)$ be the indicator function of $C^{(\epsilon)}$. Then $I_{C^{(\epsilon)}}$ is zero on the dense open subset $[0,1] \setminus C^{(\epsilon)}$ of $[0,1]$. Since $C^{(\epsilon)}$ has an empty interior, $\int_0^1 I_{C^{(\epsilon)}}(x)\,\mathrm{d}x = 0$. On the other hand, if D is a dissection of $[0,1]$, with intervals $I_1,\cdots,I_k$, and if $G = \{j : I_j \cap C^{(\epsilon)} = \varnothing\}$, then $\sum_{j \in G} l(I_j) \leqslant \epsilon$, and so $S_D(I_{C^{(s)}}) \geqslant 1 - \epsilon$. Thus $\overline{\int_0^1} I_{C^{(s)}}(x)\,\mathrm{d}x \geqslant 1 - \epsilon$.

SENTENCE STRUCTURE ANALYSIS

1. It is useful to make definitions for a general set A, but the reader should have in mind examples such as an open interval.

对一般集合 **A** 进行定义是很有用的，但是读者应该想到有一些像开区间这样的例子。

该句是一个由 but 连接的并列句。为了考虑句子平衡，第一个句子"It is useful to make definitions for a general set A"中，it 作形式主语，真正的主语是动词不定式 to make …。同样的原因，第二个句子"but the reader should have in mind examples such as an open interval"将 should have 的宾语 examples such as an open interval 放在 in mind 之后。

注:短语 have sth. in mind 记住，考虑到，想到；

bear/keep sb/sth in mind 将……记在心中；记起；考虑。

2. Since the definition of limit is so similar to that of the limit of a sequence, the proofs are simple modifications of the proofs of corresponding results for sequences.

由于极限的定义与数列极限的定义是如此相似，所以其证明是对数列相应结果的证明的简单修改。

这是一个主从复合句。其中 Since the definition of limit is so similar to that of the limit of a sequence 是原因状语从句；the proofs are simple modifications of the proofs of corresponding results for sequences 是主句。

注：表达原因的连接词或短语还有 because, because of, as, due to, owing to, on account of 等；

表达结果的连接词或短语有 so, consequently, in consequence, hence, thus, therefore, accordingly 等。我们要学会合理应用，以免在同一篇文章中重复使用。

be similar to sb/sth 与……相像的；相仿的；类似的。

Eg. The proof of Lemma 2 is similar to that of Lemma 1.

引理 2 的证明和引理 1 的相似。

3. *Suppose that* $(D_r)_{r=1}^{\infty}$ *is a sequence of dissections of* $[a,b]$, *and that* $\delta(D_r) \to 0$ *as* $r \to \infty$. *If f is a bounded function on* $[a,b]$, *then f is Riemann integrable if and only if* $S_{D_r} - s_{D_r} \to 0$ *as* $r \to \infty$.

假设 $(D_r)_{r=1}^{\infty}$ 是区间 $[a,b]$ 的一个分割序列,当 $r \to \infty$ 时,$\delta(D_r) \to 0$。如果 f 是区间 $[a,b]$ 上的一个有界函数,那么当且仅当 $r \to \infty$ 得到 $S_{D_r} - s_{D_r} \to 0$ 时,则 f 是黎曼可积的。

注:suppose that … 如果,假设

类似的表达还有:if, suppose, assume (that), assuming (that), provided/providing (that)。

Eg.

(1) Suppose the lemma were false, then we could…

假设此引理不为真,那么我们可以发现……

(2) Assume the formula holds for degree K; we will prove it for $K+1$.

假设公式对度为 K 时成立,我们将证明度为 $K+1$ 时公式仍然是成立的。

WORDS AND PHRASES

TEXT A

limit 极限

sequence 序列

real-valued function 实值函数

open interval 开区间

tend to 趋于

limit point 极限点

bounded 有界的

if and only if 当且仅当

Cauchy sequence 柯西序列

bounded below 有下界的

bounded above 有上界的

increasing function 递增函数

limes inferior 下极限

decreasing function 递减函数

convergence 收敛

real number 实数

non-empty subset 非空子集

closed interval 闭区间

rational number 有理数

element 元素

sandwich principle 两边夹原理

equivalent 等价的

one-sided 单侧的,单边的

lower bound 下界

upper and lower limits 上下极限

limes superior 上极限

oscillate 振荡,振动

TEXT B

Riemann integrable 黎曼可积的

upper integral 上积分

lower integral 下积分

integrand 被积函数，被积式

elementary integral 初等积分

bounded function 有界函数

step function 阶梯函数

dissection 划分；剖分

oscillation 振荡，振动

partition 划分；分割

continuous function 连续函数

uniformly continuous 一致连续

integrability 可积性

irrational 无理的；无理数

irrational number 无理数

discontinuous 不连续的；间断的

rational point 有理点

finite set 有限集

non-negative 非负的

constant 常数，常量；不变的

dense 稠密的

fat Cantor set 胖康托尔集，史密斯-沃尔泰拉-康托集

indicator function 指示函数，示性函数

LESSON THREE

TEXT A

Convergence Results for Fourier Series

To any function $f: [0,1] \rightarrow \mathbb{C}$ that is integrable (in the sense of Lebesgue) we assign its Fourier coefficients $\{\hat{f}(k)\}_{k \in \mathbb{Z}}$ by the prescription

$$\hat{f}(k) = \int_0^1 f(x) e^{-2\pi i k x} \, dx, \quad k \in \mathbb{Z}. \tag{1}$$

The series

$$\sum_{k \in \mathbb{Z}} \hat{f}(k) e^{2\pi i k x}$$

is termed the Fourier series of f, and the convention—always in force throughout the book—is to define it as the limit, in $\mathbb{C}$, of its symmetric partial sums

$$(S_n f)(x) = \sum_{k=-n}^{n} \hat{f}(k) e^{2\pi i k x}, \quad n \geqslant 0. \tag{2}$$

Note that the general partial sums of the two-sided infinite Fourier series are of the form

$$(S_{m,n} f)(x) = \sum_{k=m}^{n} \hat{f}(k) e^{2\pi i k x},$$

with $m < n$ two integers, and allowing for $m \rightarrow -\infty$ and $n \rightarrow \infty$ independently might seem appropriate. A compelling reason to rely on symmetric sums is the following: for $k \neq 0$, the kth Fourier coefficient $\hat{f}(k)$ and the kth term of the (formal) Fourier series of a real-valued function f both have generally nonzero imaginary parts but grouping together the terms $\hat{f}(-|k|) e^{-2\pi i |k| x}$ and $\hat{f}(|k|) e^{2\pi i |k| x}$ produces a real term, $a_k \cos(2\pi |k| x) + b_k \sin(2\pi |k| x)$. If $f \in L^1[0,1]$ is such that $\{\hat{f}(k)\}_{k \in \mathbb{Z}} \in l^1(\mathbb{Z})$, then the convergence of the symmetric partial sums (2) at some $x = x_0 \in [0,1]$ ensures that the general partial sums $(S_{m,n} f)(x)$ will also converge at $x = x_0$. Nevertheless, even if $\{\hat{f}(k)\}_{k \in \mathbb{Z}} \in l^1(\mathbb{Z})$ reliance on the more demanding interpretation of the Fourier series is more than we bargained for. To see this, first note that it is equivalent to requiring that each of the series $\sum_{k>0} \hat{f}(k)\,e^{2\pi ikx}$ and $\sum_{k<0} \hat{f}(k)\,e^{2\pi ikx}$ converges. Since knowing $a,b \in \mathbb{C}$ amounts to knowing $a \pm b$, the more demanding interpretation asks, in addition to the convergence of (2) that

$$(S_n^* f)(x) = \sum_{k=1}^{n} (\hat{f}(k)\,e^{2\pi ikx} - \hat{f}(-k)\,e^{-2\pi ikx}),\quad n \geq 1, \tag{3}$$

also converges. The operation of passing from the Fourier series $\sum_{k\in\mathbb{Z}} \hat{f}(k)\,e^{2\pi ikx}$ to the so-called conjugate series

$$\sum_{k\in\mathbb{Z}} \{-i\,\mathrm{sgn}(k)\}\hat{f}(k)\,e^{2\pi ikx}$$

is a subtle direction of research in harmonic analysis, and (3) is, up to a multiplicative factor of i, precisely the symmetric partial sums of the conjugate series. To illustrate the difficulties, note that even if the conjugate series converges (in the conventional sense for Fourier series) at every $x \in [0,1]$, it might not originate as the Fourier series of a function $f^* \in L^1[0,1]$. On the other hand, if the conjugate series is the Fourier series of some $f^* \in L^1[0,1]$, then the more demanding definition of convergence actually conceals two functions—f and f^*. These considerations motivate the conventional wisdom to interpret a Fourier series as the limit of its symmetric partial sums (2).

A basic problem is the representation of a more or less arbitrary given function f by its Fourier series. The various meanings of 'representation' are tantamount to ways in which the Fourier series may be said to converge to f.

Many areas of mathematics owe a debt of gratitude to the long struggle to come to terms with this issue, e. g. calculus (the definition of a function and the concept of a limit), integration/measure theory, the theory of distributions, but even modern set theory and mathematical logic. In the nineteenth century, the representation was taken to mean the pointwise convergence of the Fourier series at all points $x \in [0,1]$ to the limit $f(x)$. With the passage of time it was realised that this interpretation leaves much to be desired. For example, the Fourier series of a periodic continuous function, with period 1 may diverge at some points. Actually, Carleson proved that the Fourier series of a periodic continuous function converges almost everywhere to the function, while Kahane and Katznelson showed that for any set $E \subset [0,1]$ of measure zero, there is a periodic continuous function (with period 1) whose Fourier series diverges on E. To proceed, one option is to identify tangible classes of functions where pointwise convergence holds. Alternatively, relaxing the demand for convergence of the Fourier series to f at all points $x \in [0,1]$ to convergence of some trigonometric series

$$\sum_{k \in \mathbb{Z}} c_k(k) e^{2\pi i k x}$$

at almost all points of $[0,1]$ in the sense that corresponds to the meaning as signed for

the Fourier series of some $f \in L^1[0,1]$, that is, $\lim_{n \to \infty} \sum_{k=-n}^{n} c_k e^{2\pi i k x} \in \mathbb{C}$ exists, presents the

advantage that such a representation exists for every measurable function $f: [0,1] \to \mathbb{C}$,
but the enthusiasm wanes in view of the inherent defect of lack of uniqueness: there are
trigonometric series which converge to zero almost everywhere and which nevertheless
have at least one nonvanishing coefficient $c_k \in \mathbb{C}$. This surprising result means that a
pointwise representation of a trigonometric series at almost all points is never unique. On
the other hand, a convergent trigonometric series at all points is necessarily unique, if it
exists at all. But unfortunately there are continuous periodic functions that do not admit
such a representation. Consequently, using trigonometric series in general does not
resolve the limitations of Fourier series, since the alternative of coming to terms with
representations that are by no means unique has to be rejected on grounds of utility.
These considerations motivate the preference for Fourier series, as being more effective
operationally. New types of convergence actually justify the installing of Fourier series at
the forefront. For example, if pointwise convergence is replaced by convergence in
$L^2[0,1]$, the situation improves: the Fourier series of a function $f \in L^2[0,1]$ converges
in $L^2[0,1]$ to that function, as a consequence of the Fourier series theorem applied to
the orthonormal basis $\{e^{2\pi i k x}\}_{k \in \mathbb{Z}}$ of the Hilbert space $L^2[0,1]$. Other natural questions
arise. For example, what happens if $f \in L^1[0,1]$? Fourier series is an important tool in
the study of the differential equations of mathematical physics. When is it legitimate to
termwise differentiate a Fourier series representation? This problem leads to the study of
a new class of function spaces—Sobolev spaces. By construction, within these spaces,
one can express differentiation in a transparent way as an operation on the Fourier
coefficients. This compatibility and the fact that certain Sobolev spaces have a Hilbert
space structure begin to explain why these spaces are prominent in the study of partial
differential equations—more than spaces of periodic, continuously differentiable
functions. Furthermore, one can resort to a new interpretation that leads to the concept
of distribution—a major driving force behind the development of modern analysis. While
the level at which this book is written precludes an in-depth study of these topics, we
will nevertheless present some basic results that offer a perspective, and we will suggest

material for readers willing to embark on a more challenging program. Rather than attempt an inventory of sharp results, we try to convey the flavor of the theory of Fourier series and to impart working knowledge. With this objective in mind, we aim to balance the need to develop some intuitive grasp with the requisite of some essential techniques. While the subtleties can only be illustrated by addressing some rather technical points and by providing a solid base of examples, we will try not to get carried away and lose sight of the core. This means that some interesting aspects will not be pursued at all.

Throughout this chapter it is convenient to think of functions in $L^1[0,1]$ or in $L^2[0,1]$ as restrictions to $[0,1]$ of periodic functions $f:[0,1]\to\mathbb{C}$. Alternatively, we can extend $f:[0,1]\to\mathbb{C}$ to a periodic function $f:\mathbb{R}\to\mathbb{C}$ by setting $f(x+k)=f(x)$ for all $k\in\mathbb{Z}$. This procedure might require us to alter the value of $f(1)$ to $f(0)$, but modifications on sets of measure zero are irrelevant for Lebesgue integrals. When considering smoother classes of functions (for instance, functions that are Hölder continuous or differentiable) we tacitly assume that they are periodic.

TEXT B

Triangularizability and Schur's Lemma

We have discussed two different canonical forms for similarity: the rational canonical form, which applies in all cases and the Jordan canonical form, which applies only when the base field is algebraically closed. Let us now drop the rather strict requirements of canonical forms and look at two classes of matrices that are too large to be canonical forms (the upper triangular matrices and the almost upper triangular matrices) and a class of matrices that is too small to be a canonical form (the diagonal matrices).

The upper triangular matrices (or lower triangular matrices) have some nice properties and it is of interest to know when an arbitrary matrix is similar to a triangular matrix. We confine our attention to upper triangular matrices, since there are direct analogs for lower triangular matrices as well.

It will be convenient to make the following, somewhat nonstandard, definition.

Definition 1. *A linear operator τ on V is* **upper triangular** *with respect to an ordered basis $\mathcal{B}=(v_1,\cdots,v_n)$ if the matrix $[\tau]_{\mathcal{B}}$ is upper triangular, that is, if for all $i=1,\cdots,n$*

$$\tau(v_i) \in \langle v_1, \cdots, v_i \rangle$$

The operator τ is **upper triangularizable** if there is an ordered basis with respect to which τ is upper triangular.

As we will see next, when the base field is algebraically closed, all operators are upper triangularizable. However, since two distinct upper triangular matrices can be similar, the class of upper triangular matrices is not a canonical form for similarity. Simply put, there are just too many upper triangular matrices.

Theorem 1 (Schur's Lemma). *Let V be a finite-dimensional vector space over a field F.*

(1) *If $\tau \in \mathcal{L}(V)$ has the property that its characteristic polynomial $C_\tau(x)$ splits over F, then τ is upper triangularizable.*

(2) *If F is algebraically closed, then all operators are upper triangularizable.*

Proof. Part (2) follows from part (1). The proof of part (1) is most easily accomplished by matrix means, namely, we prove that every square matrix $A \in M_n(F)$ whose characteristic polynomial splits over F is similar to an upper triangular matrix. If $n = 1$, there is nothing to prove since all 1×1 matrices are upper triangular. Assume the result is true for $n - 1$ and let $A \in M_n(F)$. Let v_1 be an eigenvector associated with the eigenvalue $\lambda_1 \in F$ of A and extend $\{v_1\}$ to an ordered basis $(v_1, \cdots, v_n)$ for $\mathbb{R}^n$. The matrix of A with respect to $\mathcal{B}$ has the form

$$[A]_\mathcal{B} = \begin{bmatrix} \lambda_1 & * \\ 0 & A_1 \end{bmatrix}_{\text{block}}$$

for some $A_1 \in M_{n-1}(F)$. Since $[A]_\mathcal{B}$ and A are similar, we have

$$\det(xI - A) = \det(xI - [A]_\mathcal{B}) = (x - \lambda_1)\det(xI - A_1)$$

Hence, the characteristic polynomial of A_1 also splits over F, and by the induction hypothesis, there exists an invertible matrix $P \in M_{n-1}(F)$ for which $U = PA_1P^{-1}$ is upper triangular. Hence, if

$$Q = \begin{bmatrix} 1 & 0 \\ 0 & P \end{bmatrix}_{\text{block}}$$

then Q is invertible and

$$Q[A]_\mathcal{B}Q^{-1} = \begin{bmatrix} 1 & 0 \\ 0 & P \end{bmatrix}\begin{bmatrix} \lambda_1 & * \\ 0 & A_1 \end{bmatrix}\begin{bmatrix} 1 & 0 \\ 0 & P^{-1} \end{bmatrix} = \begin{bmatrix} \lambda_1 & * \\ 0 & U \end{bmatrix}$$

is upper triangular. This completes the proof.

When the base field is $F = \mathbb{R}$, not all operators are triangularizable. We can, however, achieve a form that is close to triangular. For the sake of the exposition, we make the following nonstandard definition (that is, the reader should not expect to find this definition in other books).

Definition 2. *A matrix $A \in M_n(F)$ is **almost upper triangular** if it has the form*

$$A = \begin{bmatrix} A_1 & & & \\ & A_2 & & \\ & & \ddots & \\ 0 & & & A_k \end{bmatrix}_{block}$$

where each matrix A_i either has size 1×1 or has size 2×2 with an irreducible characteristic polynomial. A linear operator $\tau \in \mathcal{L}(V)$ is almost upper triangularizable if there is an ordered basis $\mathcal{B}$ for which $[\tau]_\mathcal{B}$ is almost upper triangular.

We will prove that every real linear operator is almost upper triangularizable. In the case of a complex vector space V, any complex linear operator $\tau \in \mathcal{L}(V)$ has an eigenvalue and hence V contains a one-dimensional τ-invariant subspace. The analog for the real case is that for any real linear operator $\tau \in \mathcal{L}(V)$, the vector space V contains either a one-dimensional or a "nonreducible" two-dimensional τ-invariant subspace.

Theorem 2. *Let $\tau \in \mathcal{L}(V)$ be a real linear operator. Then V contains at least one of the following:*

(1) A one-dimensional τ-invariant subspace,

(2) A two-dimensional τ-invariant subspace W for which $\sigma = \tau/W$ has the property that $m_\sigma(x) = C_\sigma(x)$ is an irreducible quadratic. Hence, W is not the direct sum of two one-dimensional τ-invariant subspaces.

Proof. The minimal polynomial $m_\tau(x)$ of τ factors into a product of linear and quadratic factors over $\mathbb{R}$. If there is a linear factor $x - \lambda$, then λ is an eigenvalue for τ and if $\tau v = \lambda v$, then $\langle v \rangle$ is the desired one-dimensional τ-invariant subspace.

Otherwise, let $p(x) = x^2 + ax + b$ be an irreducible quadratic factor of $m_\tau(x)$ and write

$$m_\tau(x) = p(x)q(x).$$

Since $q(\tau) \neq 0$, we may choose a nonzero vector $v \in V$ such that $q(\tau)v \neq 0$.

Let

$$W = \langle q(\tau)v, \tau q(\tau)v \rangle.$$

This subspace is τ-invariant, for we have $[\tau q(\tau)v] \in W$ and

$$\tau[\tau q(\tau)v] = \tau^2 q(\tau)v = -(a\tau + b)q(\tau)v \in W$$

Hence, $\sigma = \tau | w$ is a linear operator on W. Also,

$$p(\tau) W = \{0\}$$

and so σ has a minimal polynomial dividing $p(x)$. But since $p(x)$ is irreducible and monic, $m_\sigma(x) = p(x)$ is quadratic. It follows that W is two-dimensional, for if

$$aq(\tau)v + b\tau q(\tau)v = 0$$

then $a + b\tau = 0$ on W, which is not the case. Finally, the characteristic polynomial $C_\sigma(x)$ has degree 2 and is divisible by $m_\sigma(x)$, whence $C_\sigma(x) = m_\sigma(x) = p(x)$ is irreducible. Thus, W satisfies condition (2).

Now we can prove Schur's lemma for real operators.

Theorem 3 (Schur's lemma : real case). *Every real linear operator $\tau \in \mathcal{L}(V)$ is almost upper triangularizable.*

Proof. As with the complex case, it is simpler to proceed using matrices, by showing that any $n \times n$ real matrix A is similar to an almost upper triangular matrix. The result is clear for $n = 1$ or if A is the zero matrix.

For $n = 2$, the characteristic polynomial $C(x)$ of A has degree 2 and is divisible by the minimal polynomial $m(x)$. If $m(x) = x - \lambda$ is linear, then $A = \lambda I_2$ is diagonal. If $m(x) = (x - \lambda)^2$, then A is similar to an upper triangular matrix with diagonal elements λ and if $m(x) = (x - \lambda)(x - \mu)$ with $\lambda \neq \mu$, then A is similar to a diagonal matrix with diagonal entries λ and μ. Finally, if $m(x) = C(x)$ is irreducible, then the result still holds.

Assume for the purposes of induction that any square matrix of size less than $n \times n$ is almost upper triangularizable. We wish to show that the same is true for any $n \times n$ matrix A. We may assume that $n \geqslant 3$.

If A has an eigenvector $v_1 \in \mathbb{R}^n$, then let $W = \langle v_1 \rangle$. If not, then according to Theorem 2, there is a pair of vectors $u_1, u_2 \in \mathbb{R}^n$ for which $W = \langle w_1, w_2 \rangle$ is two-dimensional and A-invariant and the characteristic and minimal polynomials of $\tau_A | w$ are equal and irreducible. Let U be a complement of W. If $\dim(W) = 1$, then let $\mathcal{B} = (u_1, u_2, \cdots, u_{n-1})$ be an ordered basis for $\mathbb{R}^n$ and if $\dim(W) = 2$, then let $\mathcal{B} = (w_1, w_2, u_1, \cdots, u_{n-2})$ be an ordered basis for $\mathbb{R}^n$. In either case, A is similar to a matrix of the form

$$[A]_\mathcal{B} = \begin{bmatrix} B_1 & * \\ 0 & A_1 \end{bmatrix}_{\text{block}}$$

where B_1 has size 1×1 or B_1 has size 2×2, with irreducible quadratic minimal polynomial. Also, A_1 has size $k \times k$, where $k = n - 1$ or $k = n - 2$.

Hence, the induction hypothesis applies to A_1 and there exists an invertible matrix $P \in M_k$ for which

$$U = PA_1P^{-1}$$

is almost upper triangular. Hence, if

$$Q = \begin{bmatrix} I_{n-k} & 0 \\ 0 & P \end{bmatrix}_{block}$$

then Q is invertible and

$$Q[A]_B Q^{-1} = \begin{bmatrix} I_{n-k} & 0 \\ 0 & P \end{bmatrix}\begin{bmatrix} B_1 & * \\ 0 & A_1 \end{bmatrix}\begin{bmatrix} I_{n-k} & 0 \\ 0 & P^{-1} \end{bmatrix} = \begin{bmatrix} B_1 & * \\ 0 & U \end{bmatrix}$$

is almost upper triangular. This completes the proof.

SENTENCE STRUCTURE ANALYSIS

1. Consequently, using trigonometric series in general does not resolve the limitations of Fourier series, since the alternative of coming to terms with representations that are by no means unique has to be rejected on grounds of utility.

因此，一般情况下使用三角级数并不能解决傅立叶级数的局限性，因为出于实用性的考虑，那些绝不是唯一的表示形式都不予考虑。

这是一个复合句。"using trigonometric series in general does not resolve the limitations of Fourier series"是主句，since 引导原因状语从句。此从句中，定语从句 that are by no means unique 修饰先行词 representations。

注：by no means 绝不；一点也不

by means of 借助……手段；依靠……方法

ground [usually pl.] a good or true reason for saying, doing or believing sth 充分的理由；根据

on the grounds that-clause 由于，以……为理由

Eg.

The case was dismissed on the grounds that there was not enough evidence.

此案以缺乏足够的证据为由被驳回。

on grounds of 根据，以……为理由

Eg.

Employers cannot discriminate on grounds of age.

雇主不得有年龄歧视。

2. Alternatively, relaxing the demand for convergence of the Fourier series to f at all points $x \in [0,1]$ to convergence of some trigonometric series

$$\sum_{k \in \mathbb{Z}} c_k e^{2\pi ikx}$$

at almost all points of $[0,1]$ in the sense that corresponds to the meaning as signed for the Fourier series of some $f \in L^1[0,1]$, that is, $\lim\limits_{n \to \infty} \sum\limits_{k=-n}^{n} c_k e^{2\pi ikx} \in \mathbb{C}$ exists, presents the advantage that such a representation exists for every measurable function $f : [0,1] \to \mathbb{C}$, but the enthusiasm wanes in view of the inherent defect of lack of uniqueness: there are trigonometric series which converge to zero almost everywhere and which nevertheless have at least one nonvanishing coefficient $c_k \in \mathbb{C}$.

或者，f 在所有点 $x \in [0,1]$ 的傅立叶级数的收敛要求降低到某个三角级数 $\sum\limits_{k \in \mathbb{Z}} c_k e^{2\pi ikx}$ 在 $[0,1]$ 上几乎所有的点收敛，即对于某个 $f \in L^1[0,1]$ 的傅立叶级数： $\lim\limits_{n \to \infty} \sum\limits_{k=-n}^{n} c_k e^{2\pi ikx} \in \mathbb{C}$ 是存在的。这表述的优点是对任一可测函数 $f : [0,1] \to \mathbb{C}$ 而言都存在这种表示形式，但是它具有缺乏唯一性的内在缺陷，即几乎在任何地方都存在收敛为零但至少有一个系数 $c_k \in \mathbb{C}$ 不趋于 0 的三角级数，这一点令人失望。

这句话中的 there are…进一步解释 the inherent defect of lack of uniqueness。其中两个 which 引导的定语从句都是修饰 trigonometric series。

WORDS AND PHRASES

TEXT A

Fourier series 傅立叶级数	Lebesgue 勒贝格
Fourier coefficient 傅立叶系数	partial sum 部分和
two-sided 双边的，双侧的	infinite 无穷的，无限的
integer 整数	imaginary part 虚部
operation 运算	conjugate series 共轭级数
harmonic analysis 调和分析	multiplicative factor 乘积因子
tantamount 等同的	concept 概念
measure theory 测度理论	distribution 分布
set theory 集合理论	mathematical logic 数学逻辑
pointwise 逐点的	periodic 周期的
diverge 发散	trigonometric series 三角级数
measurable function 可测函数	uniqueness 唯一性

nonvanishing 非零的

differential equation 微分方程

function space 函数空间

partial differential equation 偏微分方程

Hilbert space 希尔伯特空间

termwise 逐项

Sobolev space 索伯列夫空间

Lebesgue integral 勒贝格积分

TEXT B

triangularizability 三角可化性

canonical 标准的

algebraically closed 代数封闭的

upper triangular matrices 上三角矩阵

lower triangular matrices 下三角矩阵

operator 算子，运算符

characteristic polynomial 特征多项式

irreducible 不可约的，不可消的

quadratic 二次的，平方的；二次方程

irreducible quadratic 二次不可约方程

irreducible quadratic factor 不可约二次因式

minimal polynomial 最小多项式

Schur's Lemma 舒尔引理

Jordan canonical form 若尔当典范形

matrices 矩阵(matrix 复数)

diagonal matrices 对角矩阵

ordered 有序的

triangularizable 三角化的

invertible 可逆的

invariant subspace 不变子空间

monic 单一的，首一的

zero matrix 零矩阵

square matrix 方阵

LESSON FOUR

TEXT A

Results from Probability Theory

Since we view randomization as a key concept, we will need a few results from probability theory. There are, of course, many textbooks that contain these results. But since we will consider here only the special cases that we actually need, we can choose a simpler and more intuitive introduction to probability theory.

For a random experiment we let S denote the sample space, that is, the set of all possible outcomes of the experiment. We can restrict ourselves to the cases where S is either finite ($S = \{s_1, \cdots, s_m\}$) or countably infinite ($S = \{s_1, s_2, \cdots\}$). In the first case the corresponding index set is $I = \{1, \cdots, m\}$, and in the second case $I = \mathbb{N}$. A probability distribution p assigns to each outcome s_i for $i \in I$ a probability $p_i \geqslant 0$. The sum of all the probabilities p_i for $i \in I$ must have the value 1. An event A is a subset of the sample space S, i. e. a set of outcomes $\{s_i : I \in I_A\}$ for some $I_A \subseteq I$. The probability of an event A is denoted Prob (A) and is simply the sum of all p_i for $i \in I_A$. In particular, $\text{Prob}(\varnothing) = 0$ and $\text{Prob}(S) = 1$ for any probability distribution. Important statements concerning the probability of a union of events follow directly from this definition.

Remark 1. *A collection of events A_j for $j \in J$ is called pairwise disjoint if $A_j \cap A_{j'} = \varnothing$ whenever $j \neq j'$. For pairwise disjoint events A_j we have*

$$\text{Prob}(\bigcup_{j \in J} A_j) = \sum_{j \in J} \text{Prob}(A_j).$$

More generally, even if the events A_j are not pairwise disjoint we still have

$$\text{Prob}(\bigcup_{j \in J} A_j) \leqslant \sum_{j \in J} \text{Prob}(A_j).$$

The following images can be helpful. We can imagine a square with sides of length 1. Each outcome s_j is represented as a subregion R_i of the square with area p_i such that the regions are disjoint for distinct outcomes. Area and probability are both measures. Events are now regions of the square, and the areas of these regions are equal to the sum of the areas of the outcomes they contain. It is clear that the areas of disjoint events add, and that in the general case, the sum of the areas forms an upper bound, since there may be some "overlap" that causes some outcomes to be "double counted". Our random experiment is now equivalent to the random selection of a point in the square. If this point belongs to R_i, then the outcome is s_i.

What changes if we know that event B has occurred? All outcomes $s_i \notin B$ are now impossible and so have probability 0, while the outcomes $s_i \in B$ remain possible. So we obtain a new probability distribution q. For $s_i \notin B$, $q(s_i) = 0$. This means that the sum of all q_i for $s_i \in B$ must have the value 1. The relative probabilities of the outcomes $s_i, s_j \in B$ should not change just because we know that B has occurred, so $q_i/q_j = p_i/p_j$. Therefore, for some constant λ

$$q_i = \lambda p_i,$$

and

$$\sum_{i \in I_B} q_i = 1.$$

From this it follows that

$$\lambda = \left(\sum_{i \in I_B} q_i \right) \Big/ \left(\sum_{i \in I_B} p_i \right) = 1/\mathrm{Prob}(B).$$

So we define the conditional probability q by

$$q_i = \begin{cases} \dfrac{p_i}{\mathrm{Prob}(B)} & \text{if } s_i \in B \\ 0 & \text{otherwise.} \end{cases}$$

For an event A we obtain

$$q(A) = \sum_{i \in I_A} q_i = \sum_{i \in I_A \cap I_B} p_i/\mathrm{Prob}(B) = \frac{\mathrm{Prob}(A \cap B)}{\mathrm{Prob}(B)}.$$

For the conditional probability that A occurs under the assumption that B has occurred, the notation $\mathrm{Prob}(A \mid B)$ (read the probability of A given B) is used. So we have

$$\mathrm{Prob}(A \mid B) := \mathrm{Prob}(A \cap B)/\mathrm{Prob}(B).$$

This definition only makes sense when $\mathrm{Prob}(B) > 0$, since condition B can only occur if $\mathrm{Prob}(B) > 0$. Often the equivalent equation

$$\mathrm{Prob}(A \cap B) = \mathrm{Prob}(A \mid B) \cdot \mathrm{Prob}(B)$$

is used. This equation can be used even if $\mathrm{Prob}(B) = 0$. Although $\mathrm{Prob}(A \mid B)$ is not formally defined in this case, we still interpret $\mathrm{Prob}(A \mid B) \cdot \mathrm{Prob}(B)$ as 0.

If $\mathrm{Prob}(A \mid B) = \mathrm{Prob}(A)$, then the probability of the event A does not depend on whether or not B has occurred. In this case the events A and B are said to be independent events. This condition is equivalent to $\text{Prob}(A \cap B) = \text{Prob}(A) \cdot \text{Prob}(B)$ and also to $\text{Prob}(B|A) = \text{Prob}(B)$ if $\text{Prob}(A) > 0$ and $\text{Prob}(B) > 0$. The equation $\text{Prob}(A \cap B) = \text{Prob}(A) \cdot \text{Prob}(B)$ shows that the term independence is in fact symmetric with respect to A and B. Events A_j for $j \in J$ are called completely independent if for all $J' \subseteq J$,

$$\text{Prob}(\bigcap_{j \in J'} A_j) = \prod_{j \in J'} \text{Prob}(A_j)$$

holds.

To this point we have derived conditional probability from the probability distribution. Often we will go in the other direction. If we know the probability distribution of some statistic like income for every state, and we know the number of residents in each state (or even just the relative populations of the states), then we can determine the probability distribution for the entire country by taking the weighted sum of the regional probabilities. This idea can be carried over to probability and leads to the so-called *law of total probability*.

Theorem 1. *Let* $B_j (j \in J)$ *be a partition of the sample space S. Then*

$$\text{Prob}(A) = \sum_{j \in J} \text{Prob}(A|B_j) \cdot \text{Prob}(B_j).$$

Proof. The proof follows by simple computation.

$$\text{Prob}(A|B_j) \cdot \text{Prob}(B_j) = \text{Prob}_{j \in J}(A \cap B_j)$$

and so by Remark 1 we have

$$\sum_{j \in J} \text{Prob}(A|B_j) \cdot \text{Prob}(B_j) = \sum_{j \in J} \text{Prob}(A \cap B_j)$$

$$= \text{Prob}(\bigcup_{j \in J}(A \cap B_j))$$

$$= \text{Prob}(A \cap \bigcup_{j \in J} B_j) = \text{Prob}(A).$$

Now we come to the central notion of a random variable. Formally, this is simply a function $X : S \rightarrow \mathbb{R}$ so one random variable on the sample space of all people could assign to each person his or her height; another random variable could assign each person his or her weight. But random variables are more than just functions, since every probability distribution p on the sample space S induces a probability distribution on the range of X as follows:

$$\text{Prob}(X = t) : = \text{Prob}(\{s_i | X(s_i) = t\}).$$

So the probability that X takes on the value t is simply the probability of the set of all outcomes that are mapped to t by X. While we usually can't "do calculations" on a sample space, we can do them with random variables. Before we introduce the parameters of random variables, we want to derive the definition of independent random variables from the notion of independent random events. Two random variables X and Y on the probability space (S,p) (i. e. a sample space S and a probability distribution p on that sample space) are called independent if the events

$$\{X \in A\} := \{s_i \,|\, X(s_i) \in A\} \text{ and } \{Y \in B\}$$

are independent for all $A, B \subseteq \mathbb{R}$. A set of random variables $\{X_i \,|\, i \in I\}$ is called completely independent if the events $\{X_i \in A_i\}$ for $i \in I$ are completely independent for all events $A_i \subseteq \mathbb{R}$. The set of random variables $\{X_i \,|\, i \in I\}$ is pairwise independent if for any events $A_i, A_j \subseteq \mathbb{R}$ and any $i \neq j$, the events $\{X_i \in A_i\}$ and $\{X_j \in A_j\}$ are independent.

The most important parameter of a random variable is its expected value (or mean value) $E(X)$ defined by

$$E(X) := \sum_{t \in im(X)} t \cdot \text{Prob}(X = t),$$

where $im(X)$ denotes the image of the sample space under the function X. This definition presents no problems if $im(X)$ is finite (so, in particular, whenever S is finite). For countably infinite images the infinite series in the definition above is only defined if the series converges absolutely. This will not cause us any problems, however, since when we deal with computation times, the terms of the series will all be positive and we will also allow an expected value of ∞. Average-case runtime, as defined in a previous Chapter, is an expected value where the input x with $|x| = n$ is chosen randomly. For the expected runtime of a randomized algorithm the input x is fixed and the expected value is taken with respect to the random bits that are used by the algorithm. Since we have defined conditional probability, we can also talk about conditional expected value $E(X|A)$ with respect to the conditional probability $\text{Prob}(X = t|A)$. Expected value allows for easy computations.

Remark 2. *If X is a $0-1$ random variable (i. e. $im(X)$ is the set $\{0,1\}$), then*
$$E(X) = \text{Prob}(X = 1).$$

Proof. The claim follows directly from the definition:
$$E(X) = 0 \cdot \text{Prob}(X = 0) + 1 \cdot \text{Prob}(X = 1) = \text{Prob}(X = 1).$$

This very simple observation is extremely helpful, since it allows us to switch back and forth between probability and expected value. Furthermore, expected value is linear. This can be explained simply. If we consider the balances in the accounts of a bank's customers as random variables (based on a random selection of a customer), and every balance is reduced by a factor of 1.955 83 (to convert from German marks to euros), then the average balance will be reduced by this same factor. (Here we see a difference between theory and application, since in practice small differences occur due to the rounding of each balance to the nearest euro cent.) If two banks merge and the two accounts of each customer (perhaps with a balance of 0 for one or the other bank) are combined, then the mean balance after the merger will be the sum of the mean balances of the two banks separately. We will show that this holds in general for any random variables defined on the same probability space.

Theorem 2. *Let X and Y be random variables on the same probability space. Then*

(1) $\mathrm{E}(a \cdot X) = a \cdot \mathrm{E}(X)$ for $a \in \mathbb{R}$, *and*

(2) $\mathrm{E}(X + Y) = \mathrm{E}(X) + \mathrm{E}(Y)$.

Proof. Here we use a description of expected value that goes back to the individual outcomes of the sample space. Let (S,p) be the underlying probability space. Then

$$\mathrm{E}(X) = \sum_{i \in I} X(s_i) \cdot p_i.$$

This equation follows from the definition of $\mathrm{E}(X)$, since $\mathrm{Prob}(X = t)$ is the sum of all p_i with $X(s_i) = t$. It follows that

$$\mathrm{E}(a \cdot X) = \sum_{i \in I} (a \cdot X)(s_i) \cdot p_i = a \sum_{i \in I} X(s_i) \cdot p_i = a \cdot \mathrm{E}(X).$$

And

$$\mathrm{E}(X + Y) = \sum_{i \in I} (X + Y)(s_i) \cdot p_i$$

$$= \sum_{i \in I} X(s_i) \cdot p_i + \sum_{i \in I} Y(s_i) \cdot p_i$$

$$= \mathrm{E}(X) + \mathrm{E}(Y).$$

On the other hand, it is not in general the case that $\mathrm{E}(X \cdot Y) = \mathrm{E}(X) \cdot \mathrm{E}(Y)$. This can be shown by means of a simple example. Let $S = \{s_1, s_2\}$, $p_1 = p_2 = 1/2$, $X(s_1) = 0$, $X(s_2) = 2$, and $Y = X$. Then $X \cdot Y(s_1) = 0$ and $X \cdot Y(s_2) = 4$. It follows that $\mathrm{E}(X \cdot Y) = 2$, but $\mathrm{E}(X) \cdot \mathrm{E}(Y) = 1 \cdot 1 = 1$. The reason is that in our example X and Y are not independent.

Theorem 3. *If X and Y are independent random variables on the same sample space, then*

$$\mathrm{E}(X \cdot Y) = \mathrm{E}(X) \cdot \mathrm{E}(Y).$$

Proof. As in the proof of Theorem 2, we have

$$\mathrm{E}(X \cdot Y) = \sum_{i \in I} X(s_i) \cdot Y(s_i) \cdot p_i.$$

We partition I into disjoint sets $I(t,u) := \{i \mid X(s_i) = t, Y(s_i) = u\}$. From this it follows that

$$\mathrm{E}(X \cdot Y) = \sum_{t,u} \sum_{i \in I(t,u)} t \cdot u \cdot p_i = \sum_{t,u} t \cdot u \sum_{i \in I(t,u)} p_i$$

$$= \sum_{t,u} t \cdot u \cdot \mathrm{Prob}(X = t, Y = u).$$

Now we can take advantage of the independence of X and Y and obtain

$$\mathrm{E}(X \cdot Y) = \sum_{t,u} t \cdot u \cdot \mathrm{Prob}(X = t) \cdot \mathrm{Prob}(Y = u)$$

$$= \left(\sum_{t} t \cdot \mathrm{Prob}(X = t) \right) \cdot \left(\sum_{u} u \cdot \mathrm{Prob}(Y = u) \right)$$

$$= \mathrm{E}(X) \cdot \mathrm{E}(Y).$$

The claim of Theorem 3 can be illustrated as follows. If we assume that weight and account balance are independent, then the mean balance is the same for every weight, and so the mean product of account balance and weight is the product of the mean account balance and the mean weight. This example also shows that data from everyday life typically only lead to "almost independent" random variables. But we can design experiments with coin tosses so that the results are "genuinely independent".

The expected value reduces the random variable to its weighted mean, and so expresses only a portion of the information contained in a random variable and its probability distribution. The mean annual income in two countries can be the same, for example, while the income disparity in one country may be small and in the other country much larger. So we are interested in the random variable $Y = |X - \mathrm{E}(X)|$, which measures the distance of a random variable from its expected value. The kth central moment of X is $\mathrm{E}(|X - \mathrm{E}(X)|^k)$. The larger k is, the more heavily deviations from the mean are weighted. Based on the discussion above, we might expect that the first central moment would be the most important, but it is computationally inconvenient because the function $|X|$ is not differentiable. As a standard measure of deviation from the expected value, the second central moment is usually used. It is called the variance of the random variable X and denoted by $V(X) := E((X - E(X))^2)$. Since $X^2 = |X|^2$, we can drop the absolute value. Directly from the definition we obtain the following results.

Theorem 4. *Let X be a random variable such that $|E(X)| < \infty$, then*

$$V(X) = E(X^2) - E(X)^2$$

and for any $a \in \mathbb{R}$,

$$V(aX) = a^2 \cdot V(X).$$

Proof. The condition $|E(X)| < \infty$ guarantees that on the right side of the first claim we do not have the undefined quantity $\infty - \infty$. Then, since $E(X)$ is a constant factor, by linearity of expected value we have

$$V(X) = E((X - E(X))^2) = E(X^2 - 2 \cdot X \cdot E(X) + E(X)^2)$$
$$= E(X^2) - 2 \cdot E(X) \cdot E(X) + E(E(X)^2).$$

Finally, for each constant $a \in \mathbb{R}$, $E(a) = a$ since we are dealing with a 'random' variable that always takes on the value a. So $E(E(X)^2) = E(X)^2$ and we obtain the first claim. For the second claim we apply the first statement and use the equations

$$E((aX)^2) = a^2 \cdot E(X^2) \text{ and } E(aX)^2 = a^2 \cdot E(X)^2.$$

Since $V(2 \cdot X) = 4 \cdot V(X)$ and not $2 \cdot V(X)$, it is not generally the case that $V(X + Y) = V(X) + V(Y)$. This is the case, however, if the random variables are independent.

Theorem 5. *For pairwise independent random variables $X_1, \cdots, X_n$ we have*

$$V(X_1 + \cdots + X_n) = V(X_1) + \cdots + V(X_n).$$

Proof. The statement follows by simple computation. We have

$$V\left(\sum_{1 \leq i \leq n} X_i \right) = E\left(\left(\sum_{1 \leq i \leq n} X_i - E\left(\sum_{1 \leq i \leq n} X_i \right) \right)^2 \right)$$

$$= E\left(\left(\sum_{1 \leq i \leq n} X_i - \sum_{1 \leq i \leq n} E(X_i) \right)^2 \right)$$

$$= E\left(\sum_{1 \leq i,j \leq n} X_i \cdot X_j - 2 \sum_{1 \leq i,j \leq n} X_i \cdot E(X_j) + \sum_{1 \leq i,j \leq n} E(X_i) \cdot E(X_j) \right)$$

$$= \sum_{1 \leq i,j \leq n} (E(X_i \cdot X_j) - 2 \cdot E(X_i) \cdot E(X_j) + E(X_i) \cdot E(X_j)).$$

By Theorem 3 it follows from the independence of X_i and X_j, that $E(X_i, X_j) = E(X_i) \cdot E(X_j)$, and so the summands for all $i \neq j$ equal 0. For $i = j$ we obtain $E(X_i^2) - E(X_i)^2$ and thus $V(X_i)$. This proves the theorem.

TEXT B

Least Squares Method

Up to now, we have been dealing with systems of linear equations of the form $Ax = b$ where $A \in \mathbb{R}^{n \times n}$. However, it is frequently the case in practical problems (typically, in problems of data-fitting) that the matrix A is not square but rectangular, and we have to solve a linear system of equations $Ax = b$ with $A \in \mathbb{R}^{m \times n}$, $b \in \mathbb{R}^m$, with $m > n$; since there are more equations than unknowns, in general such a system will have no solution. Consider, for example, the linear system (with $m = 3, n = 2$)

$$\begin{pmatrix} 3 & 1 \\ 1 & 1 \\ 4 & 2 \end{pmatrix} \begin{pmatrix} x_1 \\ x_2 \end{pmatrix} = \begin{pmatrix} 1 \\ 0 \\ 2 \end{pmatrix};$$

by adding the first two of the three equations and comparing the result with the third, it is easily seen that there is no solution. If, on the other hand, $m < n$, then the situation is reversed and there may be an infinite number of solutions. Consider, for example, the linear system (with $m = 1, n = 2$)

$$(3 \quad 1) \begin{pmatrix} x_1 \\ x_2 \end{pmatrix} = 1;$$

any vector $x = (\mu, 1 - 3\mu)^T$, with $\mu \in \mathbb{R}$, is a solution to this system. Suppose that $m \geq n$; we may then need to find a vector $x \in \mathbb{R}^n$ which satisfies $Ax - b \approx 0$ in $\mathbb{R}^m$ as nearly as possible in some sense. This suggests that we define the residual vector $r = Ax - b$ and require to minimise a certain norm of r in $\mathbb{R}^m$. From the practical point of view, it is particularly convenient to minimise the residual vector r in the 2-norm on $\mathbb{R}^m$; this leads to the least squares problem:

$$\underset{x \in \mathbb{R}^n}{\text{Minimise}} \|Ax - b\|_2.$$

This is clearly equivalent to minimising the square of the norm; so, on noting that

$$\|Ax - b\|_2^2 = (Ax - b)^T(Ax - b),$$

the problem may be restated as

$$\underset{x \in \mathbb{R}^n}{\text{Minimise}} (Ax - b)^T(Ax - b).$$

Since

$$(Ax - b)^T(Ax - b) = x^T A^T A x - 2x^T A^T b + b^T b,$$

the quantity to be minimised is a nonnegative quadratic function of the n components of the vector $\boldsymbol{x}$; the minimum therefore exists, and may be found by equating to zero the partial derivatives with respect to the components. This leads to the system of equations

$$Bx = A^{\mathrm{T}}\boldsymbol{b}, \quad \text{where } B = A^{\mathrm{T}}A.$$

The matrix B is symmetric, and if A has full rank, n, then B is non-singular; it is called the normal matrix, and the system $Bx = A^{\mathrm{T}}\boldsymbol{b}$ is called the system of normal equations.

The normal equations have important theoretical properties, but do not lead to a satisfactory numerical algorithm, except for fairly small problems. The difficulty is that in a practical least squares problem the matrix A is likely to be quite ill-conditioned, and $B = A^{\mathrm{T}}A$ will then be extremely ill-conditioned. For example, if

$$A = \begin{pmatrix} \varepsilon & 0 \\ 0 & 1 \end{pmatrix}$$

where $\varepsilon \in (0,1)$, then $\kappa_2(A) = \varepsilon^{-1} > 1$, while

$$\kappa_2(B) = \kappa_2(A^{\mathrm{T}}A) = \varepsilon^{-2} = \varepsilon^{-1}\kappa_2(A) \gg \kappa_2(A)$$

when $0 < \varepsilon \ll 1$. If possible, one should avoid using a method which leads to such a dramatic deterioration of the condition number.

There are various alternative techniques which avoid the direct construction of the normal matrix $A^{\mathrm{T}}A$, and so do not lead to this extreme ill-conditioning. Here we shall describe just one algorithm, which begins by factorising the matrix A, but using an orthogonal matrix rather than the lower triangular factor as in the early section.

Theorem 1. *Suppose that $A \in \mathbb{R}^{m \times n}$ where $m \geq n$. Then, A can be written in the form*

$$A = \hat{Q}\hat{R},$$

where $\hat{R}$ is an upper triangular $n \times m$ matrix, and $\hat{Q}$ is an $m \times n$ matrix which satisfies

$$\hat{Q}^{\mathrm{T}}\hat{Q} = I_n, \tag{1}$$

where I_n is the $n \times n$ identity matrix; if $rank(A) = n$, then $\hat{R}$ is nonsingular.

Proof. We use induction on n, the number of columns in A. The theorem clearly holds when $n = 1$, A has only one column. Indeed, writing c for this column vector and assuming that $c \neq 0$, the matrix $\hat{Q}$ has just one column, the vector $\boldsymbol{c}/\|\boldsymbol{c}\|_2$, and $\hat{R}$ has a single element, $\|\boldsymbol{c}\|_2$. In the special case where c is the zero vector, we can choose $\hat{R}$ to have the single element 0, and $\hat{Q}$ to have a single column which can be an arbitrary vector in $\mathbb{R}^m$ whose 2-norm is equal to 1.

Suppose that the theorem is true when $n = k$, where $1 \leq k \leq m$. Consider a matrix A which has m rows and $k + 1$ columns, partitioned as

$$A = (A_k \quad \boldsymbol{a}),$$

where $\boldsymbol{a} \in \mathbb{R}^m$ is a column vector and A_k has k columns. To obtain the desired factorisation $\hat{Q}\hat{R}$ of A we seek $\hat{Q} = (\hat{Q}_k \quad \boldsymbol{q})$ and

$$\hat{R} = \begin{pmatrix} \hat{R}_k & \boldsymbol{r} \\ \mathbf{0} & \alpha \end{pmatrix}$$

such that

$$A = (A_k \quad \boldsymbol{a}) = (\hat{Q}_k \quad \boldsymbol{q}) \begin{pmatrix} \hat{R}_k & \boldsymbol{r} \\ \mathbf{0} & \alpha \end{pmatrix}.$$

Multiplying this out and requiring that $\hat{Q}^T \hat{Q} = I_{k+1}$, the identity matrix of order $k + 1$, we conclude that

$$A_k = \hat{Q}_k \hat{R}_k \tag{2}$$

$$\boldsymbol{a} = \hat{Q}_k \boldsymbol{r} + \boldsymbol{q}\alpha \tag{3}$$

$$\mathbf{Q}_k^T \hat{Q}_k = I \tag{4}$$

$$\boldsymbol{q}^T \hat{Q}_k = \mathbf{0}^T \tag{5}$$

$$\boldsymbol{q}^T \boldsymbol{q} = 1 \tag{6}$$

These equations show that $\hat{Q}_k \hat{R}_k$ is the factorisation of A_k which exists by the inductive hypothesis, and then lead to

$$\boldsymbol{r} = \hat{Q}_k^T \boldsymbol{a}$$

$$\boldsymbol{q} = (1/\alpha)(\boldsymbol{a} - \hat{Q}_k \hat{Q}_k^T \boldsymbol{a})$$

where $\alpha = \|\boldsymbol{a} - \hat{Q}_k \hat{Q}_k^T \boldsymbol{a}\|_2$. The number α is the constant required to ensure that the vector $\boldsymbol{q}$ is normalised.

The construction fails when $\boldsymbol{a} - \hat{Q}_k \hat{Q}_k^T \boldsymbol{a} = \mathbf{0}$, for then the vector $\boldsymbol{q}$ cannot be normalised. In this case we choose $\boldsymbol{q}$ to be any normalised vector in $\mathbb{R}^m$ which is orthogonal in $\mathbb{R}^m$ to all the columns of $\hat{Q}_k$, for then $\boldsymbol{q}^T \hat{Q}_k = \mathbf{0}^T$ as required. The condition at the beginning of the proof, that $k < m$, is required by the fact that when $k = m$ the matrix $\hat{Q}_m$ is a square orthogonal matrix, and there is no vector $\boldsymbol{q}$ in $\mathbb{R}^m \setminus \{\mathbf{0}\}$ such that $\boldsymbol{q}^T \hat{Q}_m = \mathbf{0}^T$.

With these definitions of $\boldsymbol{q}$, $\boldsymbol{r}$, α, $\hat{Q}_k$ and $\hat{R}_k$, we have constructed the required factors of A, showing that the theorem is true when $n = k + 1$. Since it holds when $n = 1$ the induction is complete.

Now, for the final part, suppose that $\text{rank}(A) = n$. If $\hat{R}$ were singular, there would exist a nonzero vector $p \in \mathbb{R}^n$ such that $\hat{R}p = 0$; then, $Ap = \hat{Q}\hat{R}p = 0$, and hence rank$(A) < n$, contradicting our hypothesis that rank$(A) = n$. Therefore, if rank$(A) = n$, then $\hat{R}$ is nonsingular.

The matrix factorisation whose existence is asserted in Theorem 1 is called the QR factorisation. Here, we shall present its use in the solutions to least squares problems. In subsequent related chapter we shall revisit the idea in a different context which concerns the numerical solutions to eigenvalue problems.

Theorem 2. *Suppose that $A \in \mathbb{R}^{m \times n}$, with $m \geq n$ and rank$(A) = n$, and let $b \in \mathbb{R}^m$. Then, there exists a unique least squares solution to the system of equations $Ax = b$: a vector x in $\mathbb{R}^n$ which minimizes the function $y \mapsto \|Ay - b\|_2$ over all y in $\mathbb{R}^n$. The vector x can be obtained by finding the factors $\hat{Q}$ and $\hat{R}$ of A defined in Theorem 1, and then solving the nonsingular upper triangular system $\hat{R}x = \hat{Q}^T b$.*

Proof. The matrix $\hat{Q}$ has m rows and n columns, with $m \geq n$, and it satisfies

$$\hat{Q}^T \hat{Q} = I_n.$$

We shall suppose that $m > n$, the case $m = n$ being a trivial special case with

$$x = A^{-1}b = (\hat{Q}\hat{R})^{-1}b = \hat{R}^{-1}\hat{Q}^{-1}b = \hat{R}^{-1}\hat{Q}^T b,$$

and hence $\hat{R}x = \hat{Q}^T b$, as required.

For $m > n$ now, the vector $b \in \mathbb{R}^m$ can be written as the sum of two vectors:

$$b = b_q + b_r,$$

where b_q is in the linear space spanned by the n columns of the matrix $\hat{Q}$ and b_r is in the orthogonal complement of this space in $\mathbb{R}^m$. The vector b_q is a linear combination of the columns of $\hat{Q}$, and b_r is orthogonal to every column of $\hat{Q}$; i. e. there exists $c \in \mathbb{R}^n$ such that

$$b = b_q + b_r, \quad b_q = \hat{Q}c, \quad \hat{Q}^T b_r = 0. \tag{7}$$

Now, suppose that x is the solution to $\hat{R}x = \hat{Q}^T b$, and that y is any vector in $\mathbb{R}^n$. Then,

$$\begin{aligned}
Ay - b &= \hat{Q}\hat{R}y - b \\
&= \hat{Q}\hat{R}(y - x) + \hat{Q}\hat{R}x - b \\
&= \hat{Q}\hat{R}(y - x) + \hat{Q}\hat{Q}^T b - b \\
&= \hat{Q}\hat{R}(y - x) + \hat{Q}\hat{Q}^T b_q - b_q + \hat{Q}\hat{Q}^T b_r - b_r \\
&= \hat{Q}\hat{R}(y - x) + \hat{Q}\hat{Q}^T \hat{Q}c - b_q - b_r \\
&= \hat{Q}\hat{R}(y - x) - b_r,
\end{aligned}$$

where we have used （7）repeatedly；in particular, the last equality follows by noting that $\hat{Q}^{\mathrm{T}}\hat{Q} = I_n$. Hence

$$\|A\boldsymbol{y} - \boldsymbol{b}\|_2^2$$
$$= (\boldsymbol{y} - \boldsymbol{x})^{\mathrm{T}}\hat{R}^{\mathrm{T}}\hat{Q}^{\mathrm{T}}\hat{Q}\hat{R}(\boldsymbol{y} - \boldsymbol{x}) + \boldsymbol{b}_r^{\mathrm{T}}\boldsymbol{b}_r - 2(\boldsymbol{y} - \boldsymbol{x})^{\mathrm{T}}\hat{R}^{\mathrm{T}}\hat{Q}^{\mathrm{T}}\boldsymbol{b}_r$$
$$= \|\hat{R}(\boldsymbol{y} - \boldsymbol{x})\|_2^2 + \|\boldsymbol{b}_r\|^2$$
$$\geq \|\boldsymbol{b}_r\|^2 ,$$

since $\hat{Q}^{\mathrm{T}}\boldsymbol{b}_r = \boldsymbol{0}$. Thus $\|A\boldsymbol{y} - \boldsymbol{b}\|_2$ is smallest when $\hat{R}(\boldsymbol{y} - \boldsymbol{x}) = \boldsymbol{0}$, which implies that $\boldsymbol{y} = \boldsymbol{x}$, since the matrix $\hat{R}$ is nonsingular. Hence $\boldsymbol{x}$, defined as the solution of $\hat{R}\boldsymbol{x} = \hat{Q}\boldsymbol{b}$, is the required least squares solution.

SENTENCE STRUCTURE ANALYSIS

1. The most important parameter of a random variable is its expected value （or mean value）$\mathrm{E}(X)$ defined by

$$\mathrm{E}(X) := \sum_{t \in im(X)} t \cdot \mathrm{Prob}(X = t) ,$$

where $im(X)$ denotes the image of the sample space under the function X.

随机变量最重要的参数是 $\mathrm{E}(X) := \sum_{t \in im(X)} t \cdot \mathrm{Prob}(X = t)$ 所定义的期望值（或均值）$\mathrm{E}(X)$，其中 $im(X)$ 表示函数 X 下样本空间的图像。

这句话中包含两个定语。第一个定语由过去分词短语 defined by... 作后置定语修饰 expected value （or mean value）$\mathrm{E}(X)$。第二个定语是由 where 引导的定语从句修饰前面的公式，这里 where = in which。

2. In the special case where $\boldsymbol{c}$ is the zero vector, we can choose $\hat{R}$ to have the single element 0, and $\hat{Q}$ to have a single column which can be an arbitrary vector in $\mathbb{R}^m$ whose 2-norm is equal to 1.

在 $\boldsymbol{c}$ 为零向量的特例中，我们可以选择单个元素 0 的 $\hat{R}$，并选择单列 $\hat{Q}$，该列可以是 2 范数等于 1 的 $\mathbb{R}^m$ 中的任意向量。

这句话的主干部分为 we can choose $\hat{R}$... and $\hat{Q}$...，其中 $\hat{R}$ and $\hat{Q}$ 是并列宾语。状语 In the special case where $\boldsymbol{c}$ is the zero vector 中包含定语从句 where $\boldsymbol{c}$ is the zero vector 修饰其先行词 case。在 and $\hat{Q}$ to have a single column which can be an arbitrary vector in $\mathbb{R}^m$ whose 2-norm is equal to 1 部分包含两个定语从句，其中 which 引导定

语从句修饰先行词 a single column；定语从句 whose 2-norm is equal to 1 修饰 $\mathbb{R}^m$。

WORDS AND PHRASES

TEXT A

probability 概率	probability theory 概率论
randomization 随机，随机化	random experiment 随机试验
sample space 样本空间	countably 可数地
probability distribution 概率分布	event 事件
union of events 事件并集	collection of events 事件集合
pairwise disjoint 互不相交；两两不相交	subregion 子区域
disjoint events 不相交事件	conditional probability 条件概率
assumption 假设	independent events 独立事件
weighted sum 加权和	total probability 全概率
random variable 随机变量	parameter 参数
independent random variable 独立随机变量	pairwise 两两的
expected value 期望值	mean value 平均值，均值
terms of the series 级数的项	algorithm 算法
underlying probability space 基础概率空间	coin tosses 抛硬币（coin toss 复数）
weighted mean 加权平均数	central moment 中心矩
deviation 偏差	first central moment 一阶中心矩
differentiable 可微的	second central moment 二阶中心矩
absolute value 绝对值	summand 被加数

TEXT B

least squares method 最小二乘法	data-fitting 数据拟合
rectangular 矩形的	linear system 线性系统
residual vector 残差向量	minimise（同 minimize）使最小
2-norm 二范数	partial derivative 偏导数
normal matrix 正规矩阵	numerical algorithm 数值算法
ill-conditioned 病态条件的	condition number 条件数
factorising（同 factorizing）因子分解	row 行
factorisation（同 factorization）因子分解，因式分解	
order 阶	linear combination 线性组合
normalised（同 normalized）正规化的，规范化	
unique 唯一的	orthogonal complement 正交补

LESSON FIVE

TEXT A

Differentiating Functions of a Vector Variable

We now turn to differentiation. This involves linearity: we therefore consider functions defined on a subset U of a real normed space $(E, \|\cdot\|_E)$ taking values in a real normed space $(F, \|\cdot\|_F)$. In fact, our principal concern will be with functions of several real variables (functions defined on an open subset of $\mathbb{R}^d$), but it is worth proceeding in a more general way. Firstly, this illustrates more clearly the basic ideas that lie behind the theory. Secondly, even in the case where we consider functions defined on a finite-dimensional Euclidean space, there are advantages in proceeding in a coordinate free way; not only is the notation simpler, but also the results are seen to be independent of any particular choice of coordinates.

Recall that a real-valued function f defined on an open interval I is differentiable at a point a of I if and only if there exists a real number $f'(a)$ such that if

$$r(h) = f(a+h) - f(a) - f'(a)h$$

for all non-zero h in $I - a = \{x \in \mathbb{R} : x + a \in I\}$, then $r(h)/|h| \to 0$ as $h \to 0$, that is, $r(h) = o(|h|)$. Let us set $Df_a(x) = f'(a)x$, for $x \in \mathbb{R}$. Then Df_a is a linear mapping from $\mathbb{R}$ into $\mathbb{R}$ and

$$f(a+h) = f(a) + Df_a(h) + r(h)$$

for all $h \in I - a = \{x \in \mathbb{R} : x + a \in I\}$. Thus f is differentiable at a if and only if we can write f as the sum of a constant (the value at a), a linear term $Df_a(h)$ and a small order term $r(h)$. From this point of view, differentiation is a matter of linear approximation.

These ideas extend naturally to vector-valued functions of a vector variable. Suppose that f is a function defined on an open subset U of a real normed space $(E, \|\cdot\|_E)$, taking values in a real normed space $(F, \|\cdot\|_F)$ and that $a \in U$. We say that f is differentiable at a, with derivative Df_a, if there is a continuous linear operator

$Df_a \in L(E,F)$ such that if

$$r(h) = f(a+h) - f(a) - Df_a(h)$$

for all non-zero $h \in U - a = \{x \in E : x + a \in U\}$, then $r(h)/\|h\| \to 0$ as $h \to 0$. Again, we express f as the sum of a constant (the value at a), a linear term $Df_a(h)$ and a small order term $r(h)$.

Note that we require Df_a to be a continuous linear mapping; this condition is automatically satisfied if E is finite-dimensional, since any linear operator from a finite-dimensional normed space into a normed space is continuous (guaranteed by an existing corollary). Note also that the conditions remain the same if we replace the norm on E and the norm on F by equivalent norms. In particular, when E is finite-dimensional then we can use any norm on E (and similarly for F), since any two norms on a finite-dimensional space are equivalent.

Let us consider three special cases. First, when $E = \mathbb{R}$, we set $f'(a) = Df_a(1)$ so that

$$f(a+h) = f(a) + hf'(a) + r(h) \text{ for all } h \in U - a = \{x \in E : x + a \in U\},$$

$f'(a)$ is an element of F, while Df_a is a linear mapping from $\mathbb{R}$ into F. Note that if $\|Df_a\|$ is the operator norm of Df_a, then

$$\|Df_a\| = \sup\{\|Df_a(h)\|_F : |h| \leq 1\} = \|f'(a)\|_F.$$

Secondly, suppose that H is a real Hilbert space, and that $F = \mathbb{R}$. In this case, Df_a is a continuous linear functional on H. By the Fréchet-Riesz representation theorem, linear functionals can be expressed in terms of the inner product; there exists an element ∇f_a of H such that $Df_a(h) = \langle \nabla f_a, h \rangle$. The vector ∇f_a is called the gradient of f at a. The symbol ∇ was introduced by Hamilton, and named "nabla" by Maxwell; "nabla" is the Greek word for a Hebrew harp. Nowadays, it is usually more prosaically called "grad".

Thirdly, suppose that $f = (f_1, \cdots, f_n)$ is a mapping from an open subset U of a normed space $(E, \|\cdot\|_E)$ into a finite product $(F_1, \|\cdot\|_{F_1}) \times \cdots \times (F_n, \|\cdot\|_{F_n})$ of normed spaces. Then f is differentiable at a point a of E if and only if f_j is differentiable at a for each $1 \leq j \leq n$ if so, then $Df_a = ((Df_1)_a, \cdots, (Df_n)_a)$

If f is differentiable at every point of U, we say that f is differentiable on U. If so, then $a \to Df_a$ is a mapping from U to the normed space $(L(E,F), \|\cdot\|)$ of continuous linear mappings from E into F. We say that f is continuously differentiable at a if this mapping is continuous at a, and that f is continuously differentiable on U if it is continuously differentiable at each point of U.

As a first example, if $T \in L(E, F)$, then $T(a + h) = T(a) + T(h)$ for all a and h in E, so that T is differentiable at every point of E, and $DT_a = T$.

We now have the following elementary results.

Proposition 1. *Suppose that f and g are functions defined on an open subset U of a normed space $(E, \| \cdot \|_E)$, taking values in a normed space $(F, \| \cdot \|_F)$, that $a \in U$, and that f and g are differentiable at a.*

(1) *Df_a is uniquely determined.*

(2) *If $\epsilon > 0$ there exists $\delta > 0$ such that $N_\delta(a) \subseteq U$ and such that $\|f(a + h) - f(a)\|_F \leqslant (\|Df_a\| + \epsilon) \|h\|_E$ for $\|h\|_E < \delta$.*

(3) *f is continuous at a.*

(4) *If $\lambda, \mu = \mathbb{R}$, then $\lambda f + \mu g$ is differentiable at a, with derivative $\lambda Df_a + \mu Dg_a$.*

Proof.

(1) Suppose that $T_1, T_2 \in L(E, F)$ and that

$$f(a + h) = f(a) + T_1(h) + s_1(h) = f(a) + T_2(h) + s_2(h) \text{ for } h \in U - a,$$

where $s_1(h) / \|h\|_E \to 0$ and $s_2(h) / \|h\|_E \to 0$ as $h \to 0$. Suppose that x is a non-zero element of E. Let $y = T_1(x) - T_2(x)$. Since

$$\frac{y}{\|x\|_E} = \frac{T_1(\lambda x) - T_2(\lambda x)}{\|\lambda x\|_E} = -\frac{s_1(\lambda x) - s_2(\lambda x)}{\|\lambda x\|_E} \to 0$$

as $\lambda \to 0$, $y = 0$. Since this holds for all non-zero x in E, $T_1 = T_2$.

(2) Let $f(x + h) = f(x) + Df_a(h) + r(h)$. There exists $\delta > 0$ such that $N_\delta(a) \subseteq U$ and such that $\|r(h)\|_F \leqslant \epsilon \|h\|_E$ for $\|h\|_E < \delta$. Then

$$\|f(a + h) - f(a)\|_F = \|Df_a(h) + r(h)\|_F \leqslant \|Df_a(h)\|_F + \|r(h)\|_F$$

$$\leqslant \|Df_a\| \|h\|_E + \epsilon \|h\|_E = (\|Df_a\| + \epsilon) \|h\|_E.$$

(3) Suppose that $\epsilon > 0$. Let δ satisfy the conclusions of (2), and let $\eta = \delta \epsilon / (\delta + 1)(\|Df_a\| + \epsilon)$. If $\|h\|_E < \eta$, then $\|h\|_E < \delta$, so that

$$\|f(a + h) - f(a)\|_F \leqslant (\|Df_a\| + \epsilon) \|h\|_E < \epsilon.$$

(4) As easy as in the real scalar case.

Theorem 1 (The chain rule). *Suppose that U is an open subset of a normed space $(E, \| \cdot \|_E)$ that f is a function defined on U, taking values in a normed space $(F, \| \cdot \|_F)$, and that f is differentiable at a point a of U. Suppose that V is an open set of F containing $f(U)$ and that k is a function defined on V, taking values in a normed space $(G, \| \cdot \|_G)$ and differentiable at $f(a)$. Then the function $k \circ f$ is differentiable at a, with derivative $Dk_{f(a)} \circ Df_a$.*

Proof. Let us set $b = f(a)$ and suppose that

$$f(a+h) = f(a) + Df_a(h) + r(h).$$

First we simplify the problem, by showing that we can replace k by a function j for which $Dj_b = 0$. Let $j(y) = k(y) - Dk_b(y)$, for $y \in V$. By Proposition 1, j is differentiable at b, and $Dj_b(y) = Dk_b(y) - Dk_b(y) = 0$.

Since Dk_b is linear,

$$Dk_b(f(a+h)) = Dk_b(f(a)) + Dk_b(Df_a(h)) + Dk_b(r(h)).$$

But $\|Dk_b(r(h))\|_G \leq \|Dk_b\| \|r(h)\|_F$, so that $Dk_b(r(h))/\|h\|_E \to 0$ as $h \to 0$. Thus $Dk_b \circ f$ is differentiable at a, with derivative $Dk_b \circ Df_a$. Since $(k \circ f)(x) = (j \circ f)(x) + (Dk_b \circ f)(x)$, it is therefore sufficient to show that $j \circ f$ is differentiable at a, with derivative 0. In other words, we must show that $\|j(f(a+h)) - j(f(a))\|_G / \|h\|_E \to 0$ as $h \to 0$.

Suppose that $\epsilon > 0$. Let $L = \|Df_a\| + \epsilon$. By Proposition 1 (2), there exists $\delta > 0$ such that $N_\delta(a) \subseteq U$ and $\|f(a+h) - f(a)\|_F \leq L \|h\|_E$ for $\|h\|_E < \delta$. Since j is differentiable at b, there exists $\eta > 0$ such that $N_\eta(b) \subseteq V$ and

$$\|j(b+l) - j(b)\|_G \leq \epsilon \|l\|_F / L, \text{ for } \|l\|_F < \eta.$$

If $\|h\|_E < \min(\delta, \eta/L)$, then $\|f(a+h) - f(a)\|_F < \eta$. Set $l = f(a+h) - f(a)$ then $b + l = f(a+h)$, so that

$$\|j(f(a+h)) - j(f(a))\|_G < \epsilon \|f(a+h) - f(a)\|_F / L < \epsilon \|h\|_E.$$

Since this holds for all $\epsilon > 0$, the result follows.

Corollary 1. *If f is continuously differentiable on U and k is continuously differentiable on V, then $k \circ f$ is continuously differentiable on U.*

Proof. For the functions $x \to Dk_{f(x)}$ and $x \to Df_x$ are continuous, and the composition of two continuous functions is continuous.

Let us give some examples.

Example 1. The derivative of a bilinear mapping.

Suppose that $(E, \|\cdot\|_E)$, $(F, \|\cdot\|_F)$ and $(G, \|\cdot\|_G)$ are normed spaces, and that $\|\cdot\|$ is a product norm on $E \times F$. Suppose that B is a continuous bilinear mapping from the product $E \times F$ into G. Then

$$B((x,y) + (h,k)) = B(x,y) + B(h,y) + B(x,k) + B(h,k),$$

and

$$\|B(h,k)\|_G \leq \|B\|_\infty \|h\|_E \|k\|_F \leq \|B\|_\infty \cdot \|(h,k)\|^2,$$

where $\|B\|_\infty = \sup\{\|B(x,y)\|_G : \|x\|_E \leq 1, \|y\|_F \leq 1\}$ so that $\|B(h,k)\|_G / \|(h,k)\| \to 0$

as $(h,k) \to 0$. Thus B is differentiable at each point of $E \times F$, and $DB_{(x,y)}(h,k) = B(h,y) + B(x,k)$.

Example 2. The derivative of the norm of a real inner-product space.

Suppose that E is a real inner-product space. Let $N(x) = \|x\|$. Can we differentiate N on E? If $x \in E \setminus \{0\}$, let $l_x(\lambda) = \lambda x$, then $\|l_x(\lambda)\| = |\lambda| \|x\|$, so that the mapping $N \circ l_x$ is not differentiable at 0. Consequently N is not differentiable at 0.

Suppose on the other hand that $x \neq 0$. We write N as a product of mappings, consider each factor separately, and use the chain rule. We write $N(x) = (S \circ P \circ J)(x)$, where $J:E \setminus \{0\} \to E \times E$ is defined as $J(x) = (x,x)$, $P:E \times E \to \mathbb{R}$ is the inner product map $P(x,y) = \langle x,y \rangle$ and $S:(0,\infty) \to (0,\infty)$ is the square root map $S(x) = \sqrt{x}$. Then $S(P(J(x))) = \|x\|$, and $DJ_x = J$, since J is linear,

$$DP_{(x,y)}(h,k) = \langle h,y \rangle + \langle x,k \rangle,$$

by the previous example, and $DS_x(h) = h/2\sqrt{x}$.

By the chain rule, N is differentiable at x and

$$DN_x(h) = DS_{P(J(x))} DP_{J(x)} DJ_x(h) = DS_{P(J(x))} DP_{J(x)}(h,h)$$

$$= DS_{P(J(x))}(2\langle x,h \rangle) = \frac{\langle x,h \rangle}{\|x\|}.$$

Thus N is differentiable at each point of $E \setminus \{0\}$.

TEXT B

Historical Demographical Models

1 Basic models

The most basic models deal with one single species. They provide with a convenient starting point for more general models.

Firstly, we will study the growth of a population in terms of *time* and second, its *spatial repartition*.

Let $N(t)$ be the number of individuals in the species at time t. We assume that $N(t)$ is large enough; $N(t)$ is a real number and not an integer. This assumption is not problematic when $N(t)$ is effectively huge. On the other hand, what does $N(t)$ small mean? Does it mean that $N(t)$ is (mathematically) close to zero, or that $N(t)$ is equal to some unit? We will see that some wise change of parameters usually allows to give a precise meaning to "$N(t)$ small".

The population dynamics is described by a *conservation equation*:

$$\frac{dN(t)}{dt} = births - deaths + migrations.$$

Malthusian model

In the most basic model, there are no migrations, and births and deaths are proportional to the population $N(t)$.

$$\frac{dN(t)}{dt} = \phi N(t) - \mu N(t)$$

$$= rN(t),$$

where ϕ and μ are positive constants. We can easily deduce that $N(t) = N(0)\exp(rt)$. So, if $r > 0$, the population grows exponentially. If $r < 0$, the population decays exponentially: from a biological point of view, the species disappears.

There is a borderline case when $r = 0$: population size remains constant. This solution exists from a mathematical point of view but is unrealistic from a biological point of view. This is why we avoid the study of such borderline cases in this book.

Logistic model

It seems reasonable to include the effects of the environmental resistance when the population grows in the previous model. Basically, there exists an "ideal" population size, called the *carrying capacity*. Below the carrying capacity, the population grows. Above it, it decreases. A famous paper proposed the *logistic model*:

$$\frac{dN(t)}{dt} = rN(t)\left(1 - \frac{N(t)}{K}\right) \tag{1}$$

where r and K(the carrying capacity) are two positive constants.

The logistic model is rather basic. However, in some cases its predictions might be accurate: the Belgian Pierre Verhulst, 1804 - 1849, predicted that the Belgian population would stabilize around 9. 4 millions of inhabitants, close to its current size (10. 1millions in 1994). However, Verhulst did not take into account immigration, the death toll of the wars, the drop of the birth rate... It might be a matter of chance that its prediction seems so good! Thus one should avoid coarse models like the logistic model in order to make quantitative predictions.

The equation (1) has two steady points $N = 0$ and $N = K$ where $\frac{dN(t)}{dt} = 0$. A linearization of (1) near $N = 0$ proves that the steady state $N = 0$ is unstable. The linearization near $N = K$ proves that the steady state $N = K$ is stable.

A little algebra leads to the analytic solution of (1):

$$N(t) = \frac{N(0)K\exp(rt)}{K + N(0)(\exp(rt) - 1)},$$

We can check that the population converges to K as $t \to +\infty$: K is indeed a carrying capacity.

2 General case

A general demographical model for one single species is an autonomous differential equation in the form $\frac{dN(t)}{dt} = f(N(t))$, where f is a function of N.

The steady states are given by the solutions of the equation $f(N) = 0$. Usually, 0 is a steady state since there is no spontaneous generation. Let N be a steady state. Let us find the behavior of the solution $N(t)$ near N. Set $n(t) = N(t) - N^*$. Assume that n is small enough and that f is smooth enough:

$$\frac{dn(t)}{dt} = f(N^* + n(t)) \sim n(t)f'(N^*).$$

Therefore the behavior of the solution near $N(t) = N^*$ depends on the sign of $f'(N^*)$. If $f'(N^*) < 0$, N attracts the solution $N(t)$: N^* is a stable steady state. If $f'(N^*) > 0$, the solution $N(t)$ is ejected: N^* is an unstable steady state. The global behavior of the solution can be deduced from the study of the stability of the steady state: the function $N(t)$ is monotonic, and the limit of $N(t)$ is the nearest stable steady state. Several models of this type have been studied.

Let us make a comment on the use of mathematical modeling for predictions. An important question in Ecology deals with the vanishing of a given species. One is tempted to model the dynamics of a species by a differential equation $\frac{dN(t)}{dt} = f(N(t))$ to see whether its solution $N(t)$ can vanish. We immediately see that the answer is negative if $f'(0) > 0$. It means that the conclusion ("the population cannot vanish") is contained in the mathematical model ("$f'(0) > 0$"). This example is very basic, but we always need to keep in mind that making predictions strongly depend on the *a priori* assumptions done on the model.

3 Population models with age distribution

When modeling, a first naive idea is to try to obtain a realistic model. By a realistic model, we mean a model in which no "real world assumption" has been forgotten. Two objections can be made.

- More realistic models can lead to intricate models! And such realism (of the models) has a cost: we cannot say anything on the models.

- There still remain hidden assumptions that have been forgotten. For instance, make the following experiment in a class. Firstly, study a single population with an autonomous differential equation. The qualitative conclusion is that the size of an isolated population converges to a constant. In other words, there is no oscillation or sophisticated behavior (like chaotic behavior) with one single species. Secondly, ask if some hidden assumptions have been made. Various and interesting answers are obtained, but I have never had the suggestion that working with continuous-time models instead of discrete-time models has a dramatic influence. Thirdly, study the discrete-time logistic model...

We will now study classical demographical models that lead to rather intricate situations. A drawback of the previous demographical models is that the age distribution has not been taken into account. For instance, babies can procreate as soon as they are born! Therefore, assuming delays between birth and procreation seems reasonable.

Mc Kendrick-Von Foerster equation

Let $n(t,a)$ be the population at time t and age range a. The global population at time t is $\int_0^\infty (n(t,a)\,da)$. Let $\phi(t,a)$ and $\mu(t,a)$ be the birth and death rates. During an infinitesimal time dt, $\mu(t,a)n(t,a)\,dt$ people of age a died. The birth rate only influences $n(t,0)$ (nobody is born with an age $a > 0 ...$).

The conservation equation, called Mc Kendrick-Von Foerster equation, is:

$$dn(t,a) = \frac{\partial n}{\partial t}dt + \frac{\partial n}{\partial a}da$$

$$= -\mu(t,a)n(t,a)\,dt$$

The term $\frac{\partial n}{\partial a}da$ comes from the ageing of the population. Noting that $da/dt \equiv 1$ (After one year, you are one year older!), $n(t,a)$ satisfies the following linear partial differential equation:

$$\frac{\partial n}{\partial t} + \frac{\partial n}{\partial a} = -\mu(t,a)n. \qquad (2)$$

Now we need to specify the boundary conditions. Let $n_0(a)$ be the initial age distribution:

$$n(0,a) = n_0(a). \qquad (3)$$

The other boundary condition is given by the births:

$$n(t,0) = \int_0^{+\infty} \phi(t,a)n(t,a)\,\mathrm{d}a. \qquad (4)$$

We have taken $+\infty$ as the upper limit of the age for simplicity; of course, function $a \to \phi(t,a)$ is a compactly supported function.

Resolution

We indicate a general method for solving the Mc Kendrick-Von Foerster equation when the birth and death rates are independent from the time t; they only depend on the age a.

The operator $n \to \dfrac{\partial n}{\partial t} + \dfrac{\partial n}{\partial a}$ is a linear first-order operator and there exists a change of variables that transforms the partial differential equation into an ordinary differential equation. Set:

$$\begin{cases} \xi = a, \\ \eta = t - a. \end{cases}$$

The introduction of the variable η amounts to following a generation through the time. We then have:

$$\frac{\partial n}{\partial \xi} = -\mu(\xi)n.$$

This equation is easily solved as:

$$n(\eta,\xi) = f(\eta)\exp\left\{-\int_0^\xi \mu\right\},$$

where the function f is still unknown. Let us come back to the variables (t,a):

$$n(t,a) = f(t-a)\exp\left\{-\int_0^a \mu\right\}. \qquad (5)$$

We determine the function f for negative values using the initial condition (3):

$$n(0,a) = n_0(a)$$

$$= f(-a)\exp\left\{-\int_0^a \mu\right\},$$

so:

$$f(-a) = n_0(a)\exp\left\{\int_0^a \mu\right\}. \qquad (6)$$

We calculate f for the positive values thanks to relation (4) about births.

Let:

$$L(a) = \phi(a)\exp\left\{-\int_0^a \mu\right\}.$$

The function f, for $t \geqslant 0$, satisfies:

$$f(t) = \int_0^\infty f(t-a)L(a)\,\mathrm{d}a. \qquad (7)$$

The homogeneous integral equation(7) can be solved using the Laplace transform. We first work as if the Laplace transform of f were defined.

Let:

$$\hat{f}(\lambda) = \int_0^{+\infty} e^{-\lambda t} f(t) \, dt,$$

$$\hat{L}(\lambda) = \int_0^{+\infty} e^{-\lambda t} L(t) \, dt.$$

Then:

$$\hat{f}(\lambda) = \int_0^{+\infty} L(a) \exp\{-\lambda a\} \int_0^{+\infty} f(t-a) \exp\{-\lambda(t-a)\} \, dt da$$

$$= \int_0^{+\infty} L(a) \exp\{-\lambda a\} \, da \int_{-a}^0 f(u) \exp\{-\lambda u\} \, du$$

$$+ \int_0^{+\infty} L(a) \exp\{-\lambda a\} \, da \int_0^{+\infty} f(u) \exp\{-\lambda u\} \, du,$$

and the Laplace transform of f is given by:

$$\hat{f}(\lambda)(1 - \hat{L}(\lambda)) = \int_0^{+\infty} L(a) \exp\{-\lambda a\} \, da \tag{8}$$

$$\int_{-a}^0 n_0(-u) \exp\left\{\int_0^{-u} \mu\right\} \exp(-\lambda u) \, du.$$

We give some indications on the equation (8). Consistently with their biological interpretation, the functions $\phi(a)$ and $\mu(a)$ are compactly supported:

$$L \equiv \int_0^{+\infty} L(t) \, dt < \infty.$$

Let us distinguish two cases.

(1) $L < 1$. We then have $1 - \hat{L}(\lambda) > 0$ for $\lambda \geq 0$. The function f is determined by (8).

(2) $L > 1$. There exists at least one real number λ_0 such that $\lim_{\lambda \to \lambda_0^+} \hat{f}(\lambda) = +\infty$: $f(t)$ is not bounded for $t \geq 0$ and one can expect an explosion of the population. Using the Dominated Convergence Theorem, we choose K such that $\int_0^\infty \exp(-Ka) L(a) \, da < 1$. Let $g(t) = f(t) \exp(-Kt)$ and $H(a) = L(a) \exp(-Ka)$. Equation (7) can be written as follows:

$$g(t) = \int_0^\infty g(t-a) H(a) \, da.$$

$g(t)$ has now a Laplace transform for $\lambda \geq 0$.

By the same computations we obtain:

$$\hat{g}(\lambda)(1 - \hat{H}(\lambda)) = \int_0^{+\infty} H(a)\exp\{-\lambda a\}\,\mathrm{d}a \int_{-a}^{0} n_0(-u)\exp\left\{\int_0^{-u}\mu\right\}\exp(-\lambda u)\,\mathrm{d}u.$$

$$\tag{9}$$

We have $1 - \hat{H}(\lambda) > 0$ for $\lambda \geqslant 0$ and the function $\hat{g}(\lambda)$ is determined by (9). Let us come back to $f(t) = g(t)\exp(Kt)$. This is an indication—even though not a proof—of an explosion of the population as $t \to +\infty$.

SENTENCE STRUCTURE ANALYSIS

1. Secondly, even in the case where we consider functions defined on a finite-dimensional Euclidean space, there are advantages in proceeding in a coordinate free way: not only is the notation simpler, but also the results are seen to be independent of any particular choice of coordinates.

其次，即使在有限维欧几里得空间上定义函数的情况下，以坐标自由的方式进行处理也有优点：不仅符号更简单了，而且结果被认为与任何特定的坐标选择无关。

这句话中 where 引导的定语从句修饰先行词 case，其中 defined on a finite-dimensional Euclidean space 是过去分词短语作后置定语修饰 functions。

注：not only...but also 的意思是"不但……，而且……"，也可以变形为 not only...but...as well 或者 not only...but...。使用时，应该注意：

（1）not only...but also 连接两个分句（并列句）时，如果 not only 位于第一分句主语之前，则该分句要部分倒装；后一个分句 but also 后面不用倒装。

（2）如果 Not only...but also...连接两个主语，不能使用倒装。

2. Let us set $b = f(a)$ and suppose that
$$f(a+h) = f(a) + Df_a(h) + r(h).$$
First we simplify the problem, by showing that we can replace k by a function j for which $Dj_b = 0$.

令 $b = f(a)$，假定 $f(a+h) = f(a) + Df_a(h) + r(h)$。

首先我们可以用一个满足 $Dj_b = 0$ 的函数 j 来代替 k 来简化这个问题。

注：Let us set..., and suppose that...是常用的公式推导用语。类似的如：

（1）Let M be.... Suppose/Assume/Write that .... Then...

令 M 满足…… 假定/让…… 那么……

（2）Given any $f \neq 1$, suppose that... Then...

给定任意 $f \neq 1$，假设……那么……

(3) Let P satisfy the above assumptions, then ...

令 P 满足以上的假定条件, 那么……

3. ... we always need to keep in mind that making predictions strongly depends on the *a priori* assumptions done on the model.

……我们始终需要记住, 做出预测很大程度上取决于对模型所做的先验假设。

这句话中 that 引导的从句作 keep 的宾语, 但是由于这个宾语过长, 为了考虑句子平衡, 将它放到句尾。在 that 从句中, making predictions 是主语, depend on 是谓语, the *a priori* assumptions 是宾语, done on the model 是过去分词短语作定语修饰 *a priori* assumptions。

WORDS AND PHRASES

TEXT A

differentiation 微分

finite-dimensional 有限维的

linear mapping 线性映射

derivative 导数

functional 泛函

gradient 梯度

continuously differentiable 连续可微的

bilinear mapping 双线性映射

inner-product space 内积空间

normed space 赋范空间

Euclidean space 欧几里得空间

linear approximation 线性近似

continuous linear operator 连续线性算子

inner product 内积, 数量积

finite product 有限积

chain rule 链式法则

product norm 内积范数

TEXT B

demographical model 人口统计模型

repartition 重新划分

conservation equation 守恒方程

be proportional to 成比例

borderline 临界的, 边界的

quantitative prediction 定量预测

steady state 稳态

behavior of the solution 解的性态

smooth 光滑的

global behavior 全局形态

spatial 空间的

population dynamics 种群动态

Malthusian model 马尔萨斯模型

exponentially 呈指数地

carrying capacity 承载能力

linearization 线性化

autonomous 自治的

spontaneous 自发的

attract 吸引

stability 稳定性

monotonic 单调的

isolated 孤立的

chaotic behavior 混沌行为

discrete-time model 离散时间模型

infinitesimal 无穷小的；无穷小量

initial 初始的

compactly supported function 紧支撑函数

linear first-order operator 线性一阶算子

change of variables 变量变换

initial condition 初始条件

Laplace transform 拉普拉斯变换

Dominated Convergence Theorem 控制收敛定理

a priori 先验的

converge to 收敛于

continuous-time model 连续时间模型

age distribution 年龄分布

boundary condition 边界条件

compactly 紧的

ordinary differential equation 常微分方程

homogeneous 齐次的

LESSON SIX

TEXT A

NP-Hard Problems

Now we extend our results to general computational problems, and in particular to optimization problems.

Definition 1. *A computational problem $\mathcal{P}$ is called **NP-hard** if all problems in NP polynomially reduce to $\mathcal{P}$.*

Note that the definition also applies to decision problems, and it is symmetric (in contrast to *NP*-completeness): a decision problem is *NP*-hard if and only if its complement is. *NP*-hard problems are at least as hard as the hardest problems in *NP*. But some may be harder than any problem in *NP*. A problem which polynomially reduces to some problem in *NP* is called *NP*-**easy**. A problem which is both *NP*-hard and *NP*-easy is *NP*-**equivalent**. In other words, a problem is *NP*-equivalent if and only if it is polynomially equivalent to satisfiability, where two problems and Q are called ***polynomially equivalent*** if $\mathcal{P}$ polynomially reduces to Q, and Q polynomially reduces to $\mathcal{P}$. We note:

Proposition 1. *Let P be an NP-equivalent computational problem. Then $\mathcal{P}$ has an exact polynomial-time algorithm if and only if $P = NP$.*

Of course, all *NP*-complete problems and all *coNP*-complete problems are *NP*-equivalent. Almost all problems discussed in this book are *NP*-easy since they polynomially reduce to integer programming; this is usually a trivial observation which we do not even mention.

We now formally define the type of optimization problems we are interested in:

Definition 2. *An **NP optimization problem** is a quadruple $\mathcal{P} = (X, (S_x)_{x \in X}, c,$ goal), where*

(1) X is a language over $\{0,1\}$ decidable in polynomial time;

(2) S_x is a nonempty subset of $\{0,1\}^*$ for each $x \in X$, and there exists a

polynomial p with size(y) ≤ (p)(size(x)) for all $x \in X$ and $y \in S_x$, and the language $\{(x,y):x \in X, y \in S_x\}$ is decidable in polynomial time;

(3) $c:\{(x,y):x \in X, y \in S_x\} \to Q$ is a function computable in polynomial time; and goal $\in \{\max,\min\}$.

The elements of X are called **instances** *of $\mathcal{P}$. For each instance x, the elements of S_x are called* **feasible solutions** *of x. We write $\mathrm{OPT}(x): = goal\{c(x,y):y \in S_x\}$. An* **optimum solution** *of x is a feasible solution y to x with $c(x,y) = \mathrm{OPT}(x)$.*

A **heuristic** *for $\mathcal{P}$ is an algorithm A which computes for each input $x \in X$ with $S_x \neq 0$ a feasible solution $y \in S_x$. We sometimes write $A(x): = c(x,y)$. If $A(x) = \mathrm{OPT}(x)$ for all $x \in X$ with $S_x \neq 0$, then A is an* **exact algorithm** *for $\mathcal{P}$.*

Depending on the context, $c(x,y)$ is often called the cost, the weight, the profit or the length of y. If c is nonnegative, then we say that the optimization problem has nonnegative weights. The values of c are rational numbers; we assume an encoding into binary strings as usual.

Most interesting optimization problems fall into this class, but there are some exceptions.

An optimization problem $(X, (S_x)_{x \in X}, c, goal)$ can be regarded as the computational problem $(X, \{(x,y):x \in X, y \in S_x, c(x,y) = \mathrm{OPT}(x)\})$. Hence polynomial reductions also apply to optimization problems.

Theorem 1. *Every NP optimization problem is NP-easy.*

Proof. Let $\mathcal{P} = (X, (S_x)_{x \in X}, c, goal)$ be an NP optimization problem. We polynomially reduce $\mathcal{P}$ to a decision problem $Q \in NP$. As usual we call a string $y \in \{0,1\}^p$, $p \in \mathbb{Z}_+$ lexicographically greater than a string $s \in \{0,1\}^q$, $q \in \mathbb{Z}_+$, if $y \neq s$ and $y_j > s_j$ for $j = \min\{i \in \mathbb{N}:y_i \neq s_i\}$, where $y_i: = -1$ for $i > p$ and $s_i: = -1$ for $i > q$.

If goal = max, then Q is defined as follows: Given $x \in X$, $\gamma \in \mathbb{Q}$, and $s \in \{0,1\}^*$, is there a $y \in S_x$ such that $c(x,y) \geq \gamma$ and y is equal to or lexicographically greater than s? If goal = min, then $c(x,y) \geq \gamma$ is replaced by $c(x,y) \leq \gamma$.

Observe that Q belongs to NP(y serves as certificate). We polynomially reduce $\mathcal{P}$ to Q as follows.

As c is computable in polynomial time, there is a constant $d \in \mathbb{N}$ such that $size(c(x,y)) \leq (size(x) + p(size(x)))^d = :k(x)$ for all $x \in X$ and $y \in S_x$. Hence $\mathrm{OPT}(x) \in [-2^{k(x)}, 2^{k(x)}]$, and $|c(x,y) - c(x,y')|$ is an integral multiple of $2^{-k(x)}$ for all $x \in X$ and $y,y' \in S_x$.

Given an instance $x \in X$, we first compute $k(x)$ and then determine $\mathrm{OPT}(x)$ by binary search. We start with $\alpha := -2^{k(x)}$ and $\beta := 2^{k(x)}$. In each iteration we apply the oracle to (x,γ,s_0), where $\gamma = \dfrac{\alpha+\beta}{2}$ and s_0 is the empty string. If the answer is yes, we set $\alpha := \gamma$, otherwise $\beta := \gamma$.

After $2k+2$ iterations we have $\beta - \alpha < 2^{-k(x)}$. Then we fix $\gamma := \alpha$ and use another $2p(size(x))$ oracle calls to compute a solution $y \in S_x$ with $c(x,y) \geqslant \alpha$. For $i := 1,\cdots,p$ $(size(x))$ we call the oracle to (x,α,s_{i-1}^0) and (x,α,s_{i-1}^1), where s^j results from the string s by appending the symbol $j \in \{0,1\}$. If the both answers are yes, then we set $s_i := s_{i-1}^1$, if only the first answer is yes, then we set $s_i := s_{i-1}^0$, and if both answers are no, then we set $s_i := s_{i-1}$. We conclude that $s_{p(size(x))}$ is the lexicographically maximal string y with $y \in S_x$ and $c(x,y) = OPT(x)$.

Most problems that we discuss from now on are also NP-hard, and we shall usually prove this by describing a polynomial reduction from an NP-complete problem. As a first example we consider MAX-2SAT: given an instance of satisfiability with exactly two literals per clause, find a truth assignment that maximizes the number of satisfied clauses.

Theorem 2. (Garey, Johnson and Stockmeyer (1976))

MAX-2SAT *is NP-hard.*

Proof. By reduction from 3SAT. Given an instance I of 3SAT with clauses $C_1,\cdots,C_m$, we construct an instance I' of MAX-2SAT by adding new variables $y_1,z_1,\cdots,y_m,z_m$ and replacing each clause $C_i = \{\lambda_1,\lambda_2,\lambda_3\}$ by the fourteen clauses

$$\{\lambda_1,z_i\},\{\lambda_1,\bar{z}_i\},\{\lambda_2,z_i\},\{\lambda_2,\bar{z}_i\},\{\lambda_3,z_i\},\{\lambda_3,\bar{z}_i\},\{y_i,z_i\},\{y_i,\bar{z}_i\},$$
$$\{\lambda_1,\bar{y}_i\},\{\lambda_2,\bar{y}_i\},\{\lambda_3,\bar{y}_i\},\{\bar{\lambda}_1,\bar{\lambda}_2\},\{\bar{\lambda}_1,\bar{\lambda}_3\},\{\bar{\lambda}_2,\bar{\lambda}_3\}$$

Note that no truth assignment satisfies more than 11 of these 14 clauses. Moreover, if 11 of these clauses are satisfied, then at least one of $\lambda_1,\lambda_2,\lambda_3$ must be *true*. On the other hand, if one of $\lambda_1,\lambda_2,\lambda_3$ is *true* we can set $y_i := \lambda_1 \wedge \lambda_2 \wedge \lambda_3$ and $z_i := true$ in order to satisfy 11 of these clauses.

We conclude that I has a truth assignment satisfying all m clauses if and only if I' has a truth assignment that satisfies $11m$ clauses.

It is an open question whether each NP-hard decision problem $\mathcal{P} \in NP$ is NP-complete.

Unless $P = NP$ there is no exact polynomial-time algorithm for any NP-hard problem. There might, however, be a pseudopolynomial algorithm.

Definition 3. *Let $\mathcal{P}$ be a decision problem or an optimization problem such that each* instance x consists of a list of nonnegative integers. We denote by $largest(x)$, the largest of these integers. An algorithm for $\mathcal{P}$ is called **pseudopolynomial** if its running time is bounded by a polynomial in $size(x)$ and $largest(x)$.

For example, there is a trivial pseudopolynomial algorithm for PRIME which divides the natural number n to be tested for primality by each integer from 2 to $\lfloor \sqrt{n} \rfloor$. Another example is:

Theorem 3. *There is a pseudopolynomial algorithm for Subset-Sum.*

Proof. Given an instance $c_1, \cdots, c_n, K$ of Subset-Sum, we construct a digraph G with vertex set $\{0, \cdots, n\} \times \{0, 1, 2, \cdots, K\}$. For each $j \in \{1, \cdots, n\}$ we add edges $((j-1,i),(j,i))(i=0,1,\cdots,K)$ and $((j-1,i),(j,i+c_j))(i=0,1,\cdots,K-c_j)$.

Observe that any path from $(0,0)$ to (j,i) corresponds to a subset $S \subseteq \{1, \cdots, j\}$ with $\sum_{k \in S} c_k = i$, and vice versa. Therefore we can solve our Subset-Sum instance by checking whether G contains a path from $(0,0)$ to (n,K). With the graph scanning slgorithm this can be done in $O(nK)$ time, so we have a pseudopolynomial algorithm.

The above is also a pseudopolynomial algorithm for partition because $\frac{1}{2}\sum_{i=1}^{n} c_i \leq \frac{n}{2}$ $largest(c_1, \cdots, c_n)$. We shall discuss an extension of this algorithm. If the numbers are not too large, a pseudopolynomial algorithm can be quite efficient. Therefore, the following definition is useful.

Definition 4. *For a decision problem $\mathcal{P} = (X,Y)$ or an optimization problem $\mathcal{P} = (X,(S_x)_{x \in X},c,goal)$, and a subset $X' \subseteq X$ of instances we define the **restriction** of $\mathcal{P}$ to X' by $\mathcal{P}' = (X',X' \cap Y)$ or $\mathcal{P}' = (X',(S_x)_{x \in X'},c,goal)$, respectively.*

*Let $\mathcal{P}$ be a decision or optimization problem such that each instance consists of a list of numbers. For a polynomial p, let $\mathcal{P}_p$ be the restriction of $\mathcal{P}$ to instances x consisting of nonnegative integers with $largest(x) \leq p(size(x))$. $\mathcal{P}$ is called **strongly** NP-**hard** if there is a polynomial p such that $\mathcal{P}_p$ is NP-hard. $\mathcal{P}$ is called **strongly** NP-**complete** if $\mathcal{P} \in NP$ and there is a polynomial p such that $\mathcal{P}_p$ is NP-complete.*

Proposition 2. *Unless $P = NP$ there is no exact pseudopolynomial algorithm for any strongly NP-hard problem.*

We give some famous examples:

Theorem 4. *Integer Programming is strongly NP-hard.*

Proof. For an undirected graph G the integer program $\max\{\mathbb{1}x : x \in \mathbb{Z}^{V(G)}, 0 \leq x \leq$

$1, x_v + x_w \leqslant 1$ for $\{v,w\} \in E(G)\}$ has optimum value at least k if and only if G contains a stable set of cardinality k. Since $k \leqslant |V(G)|$ for all nontrivial instances (G,K) of STABLE SET.

TRAVELING SALESMAN PROBLEM (TSP)

Instance: A complete graph $K_n (n \geqslant 3)$ and weights $c : E(K_n) \to \mathbb{R}_+$.

Task: Find a Hamiltonian circuit T whose weight $\sum_{e \in E(T)} c(e)$ is minimum.

The vertices of a TSP-instance are often called cities, the weights are also referred to as distances.

Theorem 5. *The TSP is strongly NP-hard.*

Proof. We show that the TSP is NP-hard even when restricted to instances where all distances are 1 or 2. We describe a polynomial reduction from the HAMILTONIAN CIRCUIT problem. Given a graph G on $n \geqslant 3$ vertices, we construct the following instance of TSP: take one city for each vertex of G, and let the distances be 1 whenever the edge is in $E(G)$ and 2 otherwise. It is then obvious that G is Hamiltonian if and only if the length of an optimum TSP tour is n.

The proof also shows that the following decision problem is not easier than the TSP itself: given an instance of the TSP and an integer k, is there a tour of length k or less? A similar statement is true for a large class of discrete optimization problems:

Proposition 3. *Let $\mathcal{F}$ and $\mathcal{F}'$ be (infinite) families of finite sets, and let $\mathcal{P}$ be the following optimization problem: given a set $E \in \mathcal{F}$ and a function $c : E \to \mathbb{Z}$, find a set $F \subseteq E$ with $F \in \mathcal{F}'$ and $c(F)$ minimum (or decide that no such F exists).*

Then $\mathcal{P}$ can be solved in polynomial time if and only if the following decision problem can be solved in polynomial time: given an instance (E,c) of $\mathcal{P}$ and an integer k, is $\mathrm{OPT}((E,c)) \leqslant k$? If the optimization problem is NP-hard, then so is this decision problem.

Proof. It suffices to show that there is an oracle algorithm for the optimization problem using the decision problem (the converse is trivial). Let (E,c) be an instance of $\mathcal{P}$. We first determine $\mathrm{OPT}((E,c))$ by binary search. Since there are at most $1 + \sum_{e \in E} |c(e)| \leqslant 2^{size(c)}$ possible values we can do this with $O(size(c))$ iterations, each including one oracle call.

Then we successively check for each element of E whether there exists an optimum solution without this element. This can be done by increasing its weight (say by one) and checking whether this also increases the value of an optimum solution. If so, we keep the old weight, otherwise we indeed increase the weight. After checking all elements of E, those elements whose weight we did not change constitute an optimum solution.

Examples where this result applies are the TSP, the Maximum Weight Clique Problem, the Shortest Path Problem, the Knapsack Problem, and many others. Schulz and Orlin, Punnen and Schulz proved similar results for INTEGER PROGRAMMING.

TEXT B

Combinatorial Optimization

Combinatorial optimization problems arise in several applications. Examples are the task of finding the shortest path from Paris to Rome in the road network of Europe or scheduling exams for given courses at a university. In this chapter, we give a basic introduction to the field of combinatorial optimization. Later on, we discuss how to measure the computational complexity of algorithms applied to these problems and point out some general limitations for solving difficult problems.

1　Combinatorial Optimization

Optimization problems can be divided naturally into two categories. The first category consists of problems with continuous variables. Such problems are well known from school courses on mathematics. A simple example consists of finding the minimum of the function $f: R \rightarrow R$ with $f(x) = x^2$. It is obvious that $x_0 = 0$ is the unique solution for this problem. More complicated problems are often tackled by computing the derivatives, using Newton methods or linear programming techniques. As this book deals with combinatorial optimization problems, we will not go into details the different methods to tackle continuous optimization problems, and let the interested reader refer to Nocedal and Wright.

In the case of discrete variables we are dealing with discrete optimization. When speaking of combinatorial optimization problems, most people have "natural" discrete optimization problems in mind, such as computing the shortest path or scheduling different jobs on a set of available machines. In a combinatorial optimization problem, one aims at either minimizing or maximizing a given objective function under a given set of constraints.

A problem consists of a general question that has to be answered and is given by a set of input parameters. An instance of a problem is given by the problem together with a specified parameter setting. Formally, a combinatorial optimization problem can be defined as a triple (S, f, Ω) where S is a given search space, f is the objective function, which should be either maximized or minimized, and Ω is the set of constraints that have to be fulfilled to obtain feasible solutions. The goal is to find a globally optimal solution, which is in the case of a maximization problem a solution S^* with the highest objective value that fulfills all constraints. Similarly, in the case of minimization problems, one tries to achieve a smallest objective value under the condition that all constraints are fulfilled.

Throughout this book, we consider many combinatorial optimization problems on graphs. A directed graph G is a pair $G = (V, E)$, where V is a finite set and E is a binary relation on V. The elements of V are called vertices. E is called the edge set of G and its elements are called edges. For an illustration see Figure 1.

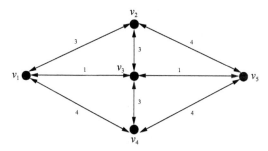

Fig. 1   Example graph G

We use the notation $e = (u, v)$ for an edge in a directed graph. Note that self-loops that are edges of the kind (u, v) are possible. In an undirected graph $G = (V, E)$, no self-loops are possible. The edge set E consists of unordered pairs of vertices in this case, and an edge is a set $\{u, v\}$ consisting of two distinct vertices $u, v \in V$. Note that one can think of an undirected edge $\{u, v\}$ as two directed edges (u, v) and (v, u). If $\{u, v\}$ is an edge in a directed graph $G = (V, E)$, we say that v is adjacent to vertex u. This leads to the representation of graphs by adjacency matrices, which will be discussed later in greater detail. A path of length k from a vertex v_0 to a vertex v_k in a graph $G = (V, E)$ is a sequence $v_0, v_1, \cdots, v_k$ of vertices such that $(v_{i-1}, v_i) \in E, 1 \leqslant i \leqslant n$, holds. Note that a path implies a sequence of directed edges. Therefore, it is sometimes useful to denote a path $(v_0, v_1, \cdots, v_k)$ by its sequence of directed edges $(v_0, v_1), (v_1, v_2), \cdots, (v_{k-1}, v_k)$.

The graph G in Figure 1 consists of the vertex set

$$V = \{v_1, v_2, v_3, v_4, v_5\}$$

and the edge set

$$E = \{e_1, e_2, e_3, e_4, e_5, e_6, e_7, e_8\}$$

where $e_1 = \{v_1, v_2\}$, $e_2 = \{v_1, v_3\}$, $e_3 = \{v_1, v_4\}$, $e_4 = \{v_2, v_3\}$, $e_5 = \{v_2, v_5\}$, $e_6 = \{v_3, v_4\}$, $e_7 = \{v_3, v_5\}$, and $e_8 = \{v_4, v_5\}$. In addition, there is a weight function $w: E \to \mathbb{N}$ assigning weights to the edges, i. e. $w(e_1) = w(e_4) = w(e_6) = 3$, $w(e_2) = w(e_7) = 1$, and $w(e_3) = w(e_5) = w(e_8) = 4$. Clearly, (v_1, v_2, v_3, v_5) is a path in G whereas (v_1, v_5, v_2) is not as there is no edge from v_1 to v_5.

There are many well-known combinatorial optimization problems on weighted graphs. We want to introduce two basic problems in the following. In the case of the single source shortest path problem, an undirected connected graph $G = (V, E)$ with positive weights on the edges is given. The goal is to compute from a designated vertex $s \in V$ the shortest paths to all other vertices of $V \backslash \{s\}$. The solution of this problem can be given by a tree rooted at s which contains the shortest paths. Considering the graph G of Figure 1 and $s = v_1$ a shortest path tree is shown in Figure 2. Another well-known combinatorial optimization problem on undirected connected graphs with positive weights is the minimum spanning tree problem. Here, one searches for a connected subgraph of the given graph G that has minimal cost. As the edge weights are positive, such a graph does not contain cycles, i. e. it is a tree. Considering again the graph G of Figure 1, a minimum spanning tree of G is given in Figure 3.

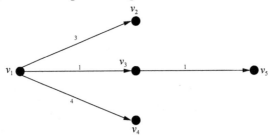

Fig. 2   Single source shortest path tree for G and $s = v_1$

Other important problems on graphs are covering problems. In the case of the so-called vertex cover problem for a given undirected graph $G = (V, E)$, one searches for a minimal subset of vertices $V' \subseteq V$ such that each edge $e \in E$ contains at least one vertex of V, i. e. $\forall e \in E: e \cap V' \neq \varnothing$ holds.

Another class of combinatorial optimization problems that has been widely examined in the literature is scheduling problems. Here, n jobs are given that have to be

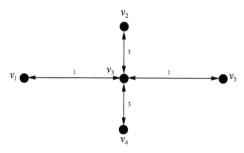

Fig. 3   Minimum spanning tree of G

processed on $m \geqslant 1$ machines. Associated with each job j, $1 \leqslant j \leqslant n$, is usually a processing time p_j. The processing time need not be the same for each machine. There are variants of scheduling problems where the processing time may depend on the machine by which it is processed. Often, also a specific due date for each job is given. Consider the following simple scheduling problem on two machines. n jobs and for each job j a processing time p_j which holds independently of the chosen machine are given. The goal is to find an assignment of the jobs to the two machines such that the overall completion time is minimized. Let $x \in \{0,1\}^n$ be a decision vector. Job j is on machine 1 iff $x_j = 0$ holds and on machine 2 iff $x_j = 1$ holds. The goal is to minimize

$$\max\Big\{ \sum_{i=1}^{n} p_j x_j, \ \sum_{i=1}^{n} p_j(1-x_j) \Big\}.$$

SENTENCE STRUCTURE ANALYSIS

1. We show that the TSP is *NP*-hard even when restricted to instances where all distances are 1 or 2.

我们证明了即使在所有距离都是 1 或 2 的情况下，TSP 也是 *NP* 难题。

这是一个 that 引导的宾语从句。此从句中，when restricted to instances 等于 when TSP is restricted to instances；定语从句 where all distances are 1 or 2 修饰其先行词 instances。

2. It suffices to show that there is an oracle algorithm for the optimization problem using the decision problem.

这足以说明 oracle 算法利用决策问题来解决优化问题。

注：suffice：to be enough for sb/sth 足够；足以。

Eg.

One example will suffice to illustrate the point.

举一个例子就足以说明这一点。

表示"足以表明/证明"的句式：

It suffices to show/prove that…

It is sufficient to show/prove that…

3. Formally, a combinatorial optimization problem can be defined as a triple (S, f, Ω) where S is a given search space, f is the objective function, which should be either maximized or minimized, and Ω is the set of constraints that have to be fulfilled to obtain feasible solutions.

一个组合优化问题在形式上可以定义为一个三元组(S,f,Ω)，其中S是一个给定的搜索空间，f是一个或最大化或最小化的目标函数，Ω是必须满足可行解的约束集。

这句话的主干部分是 a combinatorial optimization problem can be defined as a triple (S, f, Ω)；where 引导定语从句修饰 a triple (S,f,Ω)。在此定语从句中，which 引导的非限制性定语从句修饰 objective function；that 引导定语从句修饰 the set of constraints。

WORDS AND PHRASES

TEXT A

NP-hard problem　*NP* 难题(*NP*:non-deterministic polynomial 非确定性多项式)

polynomially 多项式地	reduce 归约
decision problem 决策问题	*NP*-completeness　*NP* 完全性
iff 当且仅当	satisfiability 可满足性
equivalent 等价的	polynomial-time algorithm 多项式时间算法
integer programming 整数规划	quadruple 四重的；四倍的
decidable 可判定的	polynomial time 多项式时间
goal 目标	instance 实例；样品
feasible solution 可行解	optimum solution 最优解
heuristic 启发式的	encoding 编码
binary string 二进制字符串	polynomial reduction 多项式规约
lexicographically 字典式地	integral multiple 整数倍
binary search 二分法检索；折半检索	iteration 迭代
empty string 空字符串	symbol 符号
NP-complete problem　*NP* 完全问题	truth assignment 真值分配

maximize 使最大，最大化

pseudopolynomial 伪多项式的

natural number 自然数

primality 素数性，质数性

digraph 有向图

vertex set 顶点集合

edge 边

path 路径

restriction 约束，限制

undirected graph 无向图

optimum value 最优值

complete graph 完全图

vertices (vertex 复数) 顶点

Hamiltonian 哈密尔顿算子；哈密尔顿的

family 族

TEXT B

combinatorial optimization 组合优化

categories (category 复数) 范畴

objective function 目标函数

constraint 约束

input parameter 输入参数

triple 三元组

search space 搜索空间

globally optimal solution 全局最优解

maximization problem 极大化问题

minimization problem 极小化问题

directed graph 有向图

self-loop 自圈；自循环

unordered pair 无序对

be adjacent to 邻接，邻近，毗邻

adjacency matrices 邻接矩阵

directed edge 定向边

weighted graph 加权图

minimal subset 最小子集

single source shortest path problem 单源最短路径问题

connected graph 连接图

connected subgraph 连接子图

minimum spanning tree problem 最小生成树问题

scheduling problem 调度问题

LESSON SEVEN

TEXT A

Relaxation and Newton's Method

In the previous section we saw how various ingenious devices lead to iterations which may or may not converge to the desired solutions to a given equation $f(x) = 0$. We would obviously benefit from a more generally applicable iterative method which would, except possibly in special cases, produce a sequence (x_k) that always converges to a required solution. One way of constructing such a sequence is by relaxation.

Definition 1 *Suppose that f is a real-valued function, defined and continuous in a neighbourhood of a real number ξ. Relaxation uses the sequence (x_k) defined by*

$$x_{k+1} = x_k - \lambda f(x_k), \quad k = 0,1,2,\cdots, \tag{1}$$

where $\lambda \neq 0$ is a fixed real number whose choice will be made clear below, and x_0 is a given starting value near ξ.

If the sequence (x_k) defined by (1) converges to ξ, then ξ is a solution to the equation $f(x) = 0$, as we assume that f is continuous.

It is clear from (1) that relaxation is a simple iteration of the form $x_{k+1} = g(x_k)$, $k = 0,1,2,\cdots$, with $g(x) = x - \lambda f(x)$. Suppose now, further, that f is differentiable in a neighbourhood of ξ. It then follows that $g'(x) = 1 - \lambda f'(x)$ for all x in this neighbourhood; hence, if $f(\xi) = 0$ and $f'(\xi) \neq 0$, the sequence (x_k) defined by the iteration $x_{k+1} = g(x_k)$, $k = 0,1,2,\cdots$, will converge to ξ if we choose λ to have the same sign as $f'(\xi)$, to be not too large, and take x_0 sufficiently close to ξ. This idea is made more precise in the next theorem.

Theorem 1. *Suppose that f is a real-valued function, defined and continuous in a neighbourhood of a real number ξ, and let $f(\xi) = 0$. Suppose further that f is defined and continuous in some neighbourhood of ξ, and let $f'(\xi) \neq 0$. Then, there exist positive real numbers λ and δ such that the sequence (x_k) defined by the relaxation iteration (1) converges to ξ for any x_0 in the interval $[\xi - \delta, \xi + \delta]$.*

Proof. Suppose that $f'(\xi) = \alpha$, and that α is positive. If $f'(\xi)$ is negative, the proof is similar, with appropriate changes of sign. Since f' is continuous in some neighbourhood of ξ, we can find a positive real number δ such that $f'(x) \geqslant \frac{1}{2}\alpha$ in the interval $[\xi - \delta, \xi + \delta]$. Let M be an upper bound for $f'(x)$ in this interval. Hence $M \geqslant \frac{1}{2}\alpha$. In order to fix the value of the real number λ, we begin by noting that, for any $\lambda > 0$,

$$1 - \lambda M \leqslant 1 - \lambda f'(x) \leqslant 1 - \frac{1}{2}\lambda\alpha, \quad \lambda \in [\xi - \delta, \xi + \delta].$$

We now choose λ so that these extreme values are equal and opposite, i. e. $1 - \lambda M = -\vartheta$ and $1 - \frac{1}{2}\lambda\alpha = \vartheta$ for a suitable nonnegative real number ϑ. There is a unique value of ϑ for which this holds; it is given by the formula

$$\vartheta = \frac{2M - \alpha}{2M + \alpha},$$

corresponding to

$$\lambda = \frac{4}{2M + \alpha}.$$

On defining $g(x) = x - \lambda f(x)$, we then deduce that

$$|g'(x)| \leqslant \vartheta < 1, \quad x \in [\xi - \delta, \xi + \delta]. \tag{2}$$

Thus we can apply Theorem 1 to conclude that the sequence (x_k) defined by the relaxation iteration (1) converges to ξ, provided that x_0 is in the interval $[\xi - \delta, \xi + \delta]$. The asymptotic rate of convergence of the relaxation iteration (1) to ξ is at least $-\log_{10}\vartheta$.

We can now extend the idea of relaxation by allowing λ to be a continuous function of x in a neighbourhood of ξ rather than just a constant. This suggests an iteration

$$x_{k+1} = x_k - \lambda(x_k)f(x_k), \quad k = 0, 1, 2, \cdots$$

corresponding to a simple iteration with $g(x) = x - \lambda(x)f(x)$. If the sequence (x_k) converges, the limit ξ will be a solution to $f(x) = 0$, except possibly when $\lambda(\xi) = 0$. Moreover, as we have seen, the ultimate rate of convergence is determined by $g'(\xi)$. Since $f(\xi) = 0$, it follows that $g'(\xi) = 1 - \lambda(\xi)f'(\xi)$, and (2) suggest using a function λ which makes $1 - \lambda(\xi)f'(\xi)$ small. The obvious choice is $\lambda(x) = 1/f'(x)$, and leads us to Newton's method.

Definition 2. *Newton's method for the solution to* $f(x) = 0$ *is defined by*

$$x_{k+1} = x_k - \frac{f(x_k)}{f'(x_k)}, \quad k = 0, 1, 2, \cdots \tag{3}$$

with prescribed starting value x_0. We implicitly assume in the defining formula (3) that $f'(x_k) \neq 0$ for all $k \geq 0$.

Newton's method is a simple iteration with $g(x) = x - f(x)/f'(x)$. Its geometric interpretation is illustrated in Figure 1: the tangent to the curve $y = f(x)$ at the point $(x_k, f(x_k))$ is the line with the equation $y - f(x_k) = f'(x_k)(x - x_k)$; it meets the x-axis at the point $(x_{k+1}, 0)$.

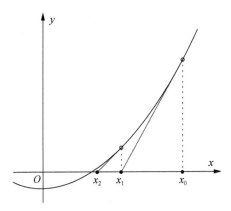

Fig. 1   Newton's method

We could apply Theorem 1 to prove the convergence of this iteration, but since generally it converges much faster than ordinary relaxation, it is better to apply a special form of proof. First, however, we give a formal definition of quadratic convergence.

Definition 3. Suppose that $\xi = \lim\limits_{k \to \infty} x_k$. We say that the sequence (x_k) converges to ξ with at least order $q > 1$, if there exists a sequence (ε_k) of positive real numbers converging to 0, and $\mu > 0$, such that

$$|x_k - \xi| \leq \varepsilon_k, \quad k = 0, 1, 2, \cdots \quad \text{and} \quad \lim_{k \to \infty} \frac{\varepsilon_{k+1}}{\varepsilon_k^q} = \mu. \tag{4}$$

If (4) holds with $\varepsilon_k = |x_k - \xi|$ for $k = 0, 1, 2, \cdots$, then the sequence (x_k) is said to converge to ξ with order q. In particular, if $q = 2$, then we say that the sequence (x_k) converges to ξ quadratically.

We note that unlike the definition of linear convergence where μ was required to belong to the interval $(0, 1)$, all we demand here is that $\mu > 0$. The reason is simple: when $q > 1$, (4) implies suitably rapid decay of the sequence (ε_k) irrespective of the size of μ.

Example 1. Let $c > 1$ and $q > 1$. The sequence (x_k) defined by $x_k = c^{-q^k}$, $k = 0, 1, 2, \cdots$, converges to 0 with order q.

Theorem 2 (Convergence of Newton's Method). *Suppose that f is a continuous real-valued function with continuous second derivative f'', defined on the closed interval $I_\delta = [\xi - \delta, \xi + \delta]$, $\delta > 0$, such that $f(\xi) = 0$ and $f''(\xi) \neq 0$. Suppose further that there exists a positive constant A such that*

$$\frac{|f''(x)|}{|f'(y)|} \leqslant A \quad \forall x, y \in I_\delta.$$

If $|\xi - x_0| \leqslant h$, where h is the smaller of δ and $1/A$, then the sequence (x_k) defined by Newton's method (3) converges quadratically to ξ.

Proof. Suppose that $|\xi - x_k| \leqslant h = \min\{\delta, 1/A\}$, so that $x_k \in I_\delta$. Then, by Taylor's Theorem, expanding about the point $x_k \in I_\delta$,

$$0 = f(\xi) = f(x_k) + (\xi - x_k)f'(x_k) + \frac{(\xi - x_k)^2}{2}f''(\eta_k), \qquad (5)$$

for some η_k between ξ and x_k, and therefore in the interval I_δ. Recalling (3), this shows that

$$\xi - x_{k+1} = -\frac{(\xi - x_k)^2 f''(\eta_k)}{2f'(x_k)}. \qquad (6)$$

Since $|\xi - x_k| \leqslant \frac{1}{A}$, we have $|\xi - x_{k+1}| \leqslant \frac{1}{2}|\xi - x_k|$. As we are given that $|\xi - x_0| \leqslant h$, it follows by induction that $|\xi - x_k| \leqslant 2^{-k}h$ for all $k \geqslant 0$; hence (x_k) converges to ξ as $k \to \infty$.

Now, η_k lies between ξ and x_k, and therefore (η_k) also converges to ξ as $k \to \infty$. Since f' and f'' are continuous on I_δ, it follows from (6) that

$$\lim_{k \to \infty} \frac{|x_{k+1} - \xi|}{|x_k - \xi|^2} = \left|\frac{f''(\xi)}{2f'(\xi)}\right|, \qquad (7)$$

which, according to Definition 2, implies quadratic convergence of the sequence (x_k) to ξ with $\mu = |f''(\xi)/2f'(\xi)|$, $\mu \in (0, A/2]$.

The conditions of the theorem implicitly require that $f'(\xi) \neq 0$, for otherwise the quantity $f''(x)/f'(y)$ could not be bounded in a neighbourhood of ξ.

One can show that if $f''(\xi) = 0$ and we assume that $f(x)$ has a continuous third derivative, and require certain quantities to be bounded, then the convergence is cubic (i. e. convergence with order $q = 3$).

It is possible to demonstrate that Newton's method converges over a wider interval, if we assume something about the signs of the derivatives.

Theorem 3. *Suppose that the function f satisfies the conditions of Theorem 1 and* *also that there exists a real number X, $X > \xi$, such that in the interval $J = [\xi, X]$ both f' and f'' are positive. Then, the sequence (x_k) defined by Newton's method (3) converges quadratically to ξ from any starting value x_0 in J.*

Proof. It follows from (6) that if $x_k \in J$, then $x_{k+1} > \xi$. Moreover, since $f'(x) > 0$ on J, f is monotonic increasing on J. As $f(\xi) = 0$, it then follows that $f(x) > 0$ for $\xi < x \leqslant X$. Hence, $\xi < x_{k+1} < x_k$, $k \geqslant 0$. Since the sequence (x_k) is bounded and monotonic decreasing, it is convergent; let $\eta = \lim_{k \to \infty} x_k$. Clearly, $\eta \in J$. Further, passing to the limit $k \to \infty$ in (3) we have that $f(\eta) = 0$. However, ξ is the only solution to $f(x) = 0$ in J, so $\eta = \xi$, and the sequence converges to ξ.

Having shown that the sequence (x_k) converges, the fact that it converges quadratically follows as in the proof of Theorem 2.

We remark that the same result holds for other possible signs of f' and f'' in a suitable interval J. The interval J does not have to be bounded; considering, for instance, $f(x) = e^x - x - 2$ from a related example, it is clear that $f'(x)$ and $f''(x)$ are both positive in the unbounded interval $(0, \infty)$, and the Newton iteration converges to the positive solution to the equation $f(x) = 0$ from any positive starting value x_0.

Note that the definition of quadratic convergence only refers to the behaviour of the sequence for sufficiently large k. In the same example we find that the convergence of the Newton iteration from a large positive value of x_0 is initially very slow. The possibility of this early behaviour is often emphasised by saying that the convergence of Newton's method is ultimately quadratic.

TEXT B

Euler's Method

Let us ponder briefly the meaning of the ordinary differential equation (ODE) ($dy = f(\tau, y(\tau)) d\tau$). We possess two items of information: we know the value of y at a single point $t = t_0$ and, given any function value $y \in \mathbb{R}^d$ and time $t \geqslant t_0$, we can tell the slope from the differential equation. The purpose of the exercise being to guess the value of y at a new point, the most elementary approach is to use linear interpolation. In other words, we estimate $y(t)$ by making the approximation $f(t, y(t)) \approx f(t_0, y(t_0))$ for $t \in [t_0, t_0 + h]$, where $h > 0$ is sufficiently small. Integrating the above ODE,

$$y(t) = y(t_0) + \int_{t_0}^{t} f(\tau, y(\tau)) d\tau \approx y_0 + (t - t_0) f(t_0, y_0). \qquad (1)$$

Given a sequence t_0, $t_1 = t_0 + h$, $t_2 = t_0 + 2h$, $\cdots$, where $h > 0$ is the time step, we denote by y_n a numerical estimate of the exact solution $y(t_n)$, $n = 0, 1, \cdots$ Motivated by (1), we choose

$$y_1 = y_0 + hf(t_0, y_0).$$

This procedure can be continued to produce approximants at t_2, t_3 and so on. In general, we obtain the recursive scheme

$$y_{n+1} = y_n + hf(t_n, y_n), \quad n = 0, 1, \cdots, \tag{2}$$

the celebrated Euler method.

Euler's method is not only the most elementary computational scheme for ODEs and, simplicity notwithstanding, of enduring practical importance, but also the cornerstone of the numerical analysis of differential equations of evolution. In a deep and profound sense, all the fancy multistep and Runge-Kutta schemes that we shall discuss are nothing but a generalization of the basic paradigm(2).

- **Graphic interpretation** Euler's method can be illustrated pictorially.

Consider, for example, the scalar *logistic equation* $y' = y(1 - y)$, $y(0) = \frac{1}{10}$. Fig. 1 displays the first few steps of Euler's method, with a grotesquely large step $h = 1$. For each step we show the exact solution with initial condition $y(t_n) = y_n$ in the vicinity of $t_n = nh$ (dotted line) and the linear interpolation via Euler's method (2) (solid line).

The initial condition being, by definition, exact, so is the slope at t_0. However, instead of following a curved trajectory the numerical solution is piecewise-linear. Having reached t_1, say, we have moved to a wrong trajectory (i. e. corresponding to a different initial condition). The slope at t_1 is wrong—or, rather, it is the correct slope of the wrong solution! Advancing further, we might well stray even more from the original trajectory.

A realistic goal of numerical solution is not, however, to avoid errors altogether; after all, we approximate it since we do not know the exact solution in the first place! An error-generating mechanism exists in every algorithm for numerical ODEs and our purpose is to understand it and to ensure that, in a given implementation, errors do not accumulate beyond a specified tolerance. Remarkably, even the excessive step $h = 1$ leads in Fig. 1 to a relatively modest local error.

Euler's method can be easily extended to cater for variable steps. Thus, for a general monotone sequence $t_0 < t_1 < t_2 < \cdots$, we approximate as follows:

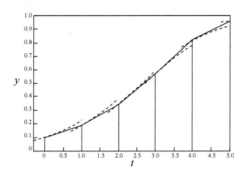

Fig. 1 Euler's method, as applied to the equation $y' = y(1 - y)$ **with initial value** $y(0) = \dfrac{1}{10}$

$$y(t_{n+1}) \approx y_{n+1} = y_n + h_n f(t_n, y_n),$$

where $h_n = t_{n+1} - t_n$, $n = 0, 1, \cdots$ However, for the time being we restrict ourselves to constant steps.

How good is Euler's method in approximating ODE? Before we even attempt to answer this question, we need to formulate it with considerably more rigour. Thus, suppose that we wish to compute a numerical solution of ODE in the compact interval $[t_0, t_0 + t^*]$ with some time-stepping numerical method, not necessarily Euler's scheme. In other words, we cover the interval by an equidistant grid and employ the time-stepping procedure to produce a numerical solution. Each grid is associated with a different numerical sequence and the critical question is whether, as $h \rightarrow 0$ and the grid is being refined, the numerical solution tends to the exact solution of ODE. More formally, we express the dependence of the numerical solution upon the step size by the notation $y_n = y_{n,h}$, $n = 0, 1, \cdots, \lfloor t^*/h \rfloor$. A method is said to be convergent if, for every ODE with a Lipschitz function f and every $t^* > 0$ it is true that

$$\lim_{h \rightarrow 0^+} \max_{n = 0, 1, \cdots, \lfloor t^*/h \rfloor} \| y_{n,h} - y(t_n) \| = 0,$$

where $\lfloor \alpha \rfloor \in \mathbb{Z}$ is the integer part of $\alpha \in \mathbb{R}$. Hence, convergence means that, for every Lipschitz function, the numerical solution tends to be the true solution as the grid becomes increasingly fine.

In the next few chapters we will mention several desirable attributes of numerical methods for ODEs. It is crucial to understand that convergence is not just another 'desirable' property but, rather, a *sine qua non* of any numerical scheme. *Unless it converges, a numerical method is useless*!

Theorem 1. *Euler's method* (2) *is convergent.*

Proof. We prove this theorem subject to the extra assumption that the function f

(and therefore also y) is analytic (it is enough, in fact, to stipulate the weaker condition of continuous differentiability).

Given $h > 0$ and $y_n = y_{n,h}$, $n = 0, 1, \cdots, \lfloor t^*/h \rfloor$, we let $e_{n,h} = y_{n,h} - y(t_n)$ denote the numerical error. Thus, we wish to prove that $\lim_{h \to 0^+} \max_n \| e_{n,h} \| = 0$.

By Taylor's theorem and the differential equation in first paragraph of this text,

$$y(t_{n+1}) = y(t_n) + hy'(t_n) + \mathcal{O}(h^2) = y(t_n) + hf(t_n, y(t_n)) + \mathcal{O}(h^2) \qquad (3)$$

and, y being continuously differentiable, the $\mathcal{O}(h^2)$ term can be bounded (in a given norm) uniformly for all $h > 0$ and $n \leqslant \lfloor t^*/h \rfloor$ by a term of the form ch^2, where $c > 0$ is a constant. We subtract (3) from (2), giving

$$e_{n+1,h} = e_{n,h} + h[f(t_n, y(t_n) + e_{n,h}) - f(t_n, y(t_n))] + \mathcal{O}(h^2).$$

Thus, it follows by the triangle inequality from the Lipschitz condition and the aforementioned bound on the $\mathcal{O}(h^2)$ reminder term that

$$\| e_{n+1,h} \| \leqslant \| e_{n,h} \| + h \| f(t_n, y(t_n) + e_{n,h}) - f(t_n, y(t_n)) \| + ch^2 \qquad (4)$$
$$\leqslant (1 + h\lambda) \| e_{n,h} \| + ch^2, \quad n = 0, 1, \cdots, \lfloor t^*/h \rfloor - 1.$$

We now claim that

$$\| e_{n,h} \| \leqslant \frac{c}{\lambda} h [(1 + h\lambda)^n - 1], \quad n = 0, 1, \cdots \qquad (5)$$

The proof is by induction on n. When $n = 0$, we need to prove that $\| e_{0,h} \| \leqslant 0$ and hence that $e_{0,h} = 0$. This is certainly true, since at t_0 the numerical solution matches the initial condition and the error is zero.

For general $n \geqslant 0$ we assume that (5) is true up to n and use (4) to argue that

$$\| e_{n+1,h} \| \leqslant (1 + h\lambda) \frac{c}{\lambda} h [(1 + h\lambda)^n - 1] + ch^2 = \frac{c}{\lambda} h [(1 + h\lambda)^{n+1} - 1].$$

This advances the inductive argument from n to $n + 1$ and proves that (5) is true. The constant $h\lambda$ is positive, therefore $1 + h\lambda < e^{h\lambda}$ and we deduce that $(1 + h\lambda)^n < e^{nh\lambda}$. The index n is allowed to range in $\{0, 1, \cdots, \lfloor t^*/h \rfloor\}$, hence $(1 + h\lambda)^n < e^{\lfloor t^*/h \rfloor h\lambda} \leqslant e^{t^*\lambda}$. Substituting into (5), we obtain the inequality

$$\| e_{n,h} \| \leqslant \frac{c}{\lambda} (e^{t^*\lambda} - 1) h, \quad n = 0, 1, \cdots, \lfloor t^*/h \rfloor.$$

Since $c(e^{t^*\lambda} - 1)/\lambda$ is independent of h, it follows that

$$\lim_{\substack{h \to 0 \\ 0 \leqslant nh \leqslant t^*}} \| e_{n,h} \| = 0.$$

In other words, Euler's method is convergent.

Euler's method can be rewritten in the form $y_{n+1} - [y_n + hf(t_n, y_n)] = 0$. Replacing

y_k by the exact solution $y(t_k)$, $k = n$, $n + 1$, and expanding the first few terms of the Taylor series about $t = t_0 + nh$, we obtain

$$y(t_{n+1}) - [y(t_n) + hf(t_n, y(t_n))]$$
$$= [y(t_n) + hy'(t_n) + \mathcal{O}(h^2)] - [y(t_n) + hy'(t_n)] = \mathcal{O}(h^2).$$

We say that the Euler's method (2) is of order 1. In general, given an arbitrary time-stepping method

$$y_{n+1} = \mathcal{Y}_n(f, h, y_0, y_1, \cdots, y_n), \quad n = 0, 1, \cdots,$$

for the ODE, we say that it is of order p if

$$y(t_{n+1}) - \mathcal{Y}_n(f, h, y(t_0), y(t_1), \cdots, y(t_n)) = \mathcal{O}(h^{p+1}),$$

for every analytic f and $n = 0, 1, \cdots$ Alternatively, a method is of *order* p if it recovers exactly every polynomial solution of degree p or less.

The order of a numerical method provides us with information about its local *behaviour*—advancing from t_n to t_{n+1}, where $h > 0$ is sufficiently small, we are incurring an error of $\mathcal{O}(h^{p+1})$. Our main interest, however, is in not the local but the global behaviour of the method: how well is it doing in a fixed bounded interval of integration as $h \to 0$. Does it converge to the true solution? How fast? Since the local error decays as $\mathcal{O}(h^{p+1})$, the number of steps increases as $\mathcal{O}(h^{-1})$. The naive expectation is that the global error decreases as $\mathcal{O}(h^p)$, but—as we will see—it cannot be taken for granted for each and every numerical method without an additional condition. As far as Euler's method is concerned, Theorem 1 demonstrates that all is well and that the error indeed decays as $\mathcal{O}(h)$.

SENTENCE STRUCTURE ANALYSIS

1. We would obviously benefit from a more generally applicable iterative method which would, except possibly in special cases, produce a sequence (x_k) that always converges to a required solution.

显然,我们将受益于一种更普遍适用的迭代方法。除了一些特殊情况,它将产生一个序列(x_k),这个序列总是收敛于所需解。

这句话包含两个定语从句:其中 which 引导的定语从句修饰 iterative method, that 引导的定语从句修饰名词 a sequence (x_k)。

2. Thus we can apply Theorem 1 to conclude that the sequence (x_k) defined by the relaxation iteration (1) converges to ξ, provided that x_0 is in the interval $[\xi - \delta, \xi + \delta]$.

因此,我们可以应用定理 1 得出由松弛迭代(1)定义的数列(x_k)收敛于ξ,前提是 x_0 在区间$[\xi-\delta,\xi+\delta]$内。

这句话的主句 Thus we can apply Theorem 1 to conclude that the sequence (x_k) defined by the relaxation iteration (1) converges to ξ 中包含一个 that 引导的宾语从句, provided that 引导条件状语从句。That 宾语从句中, 过去分词短语 defined by the relaxation iteration (1)作后置定语修饰 the sequence (x_k)。

provided (that) 如果;假设。类似的表达还有: if、suppose (that)、assume (that)、assuming (that)、providing(that)。如:

Eg.

1) Suppose g is differentiable.

假设 g 是可微的。

2) Suppose that A is a positive matrix.

假设 A 是一个正矩阵。

3) Assuming that Eq. (1)holds, we have $u = v$.

假设式(1)成立,我们得到 $u = v$。

3. We note that unlike the definition of linear convergence where μ was required to belong to the interval $(0,1)$, all we demand here is that $\mu > 0$.

我们注意到,不像线性收敛的定义要求 μ 属于区间$(0,1)$那样,我们在这里要求的是$\mu > 0$。

本句话的主干部分是 We note that all we demand here is that $\mu > 0$。这是一个宾语从句。其中 unlike the definition of linear convergence where μ was required to belong to the interval $(0,1)$是状语, where 引导的定语从句修饰 the definition of linear convergence。句子 all we demand here is that $\mu > 0$ 中, all we demand 作主语, that 引导表语从句。

WORDS AND PHRASES

TEXT A

relaxation 松弛

iterative method 迭代方法

starting value 初始值

asymptotic 渐近的

quadratic convergence 二次收敛;二阶收敛

continuous second derivative 连续二阶导数

Newton's method 牛顿法

neighbourhood 邻域

relaxation iteration 松弛迭代

tangent 切线;正切;相切的

third derivative 三阶导数
monotonic increasing 单调递增
ultimately 最终地

cubic 立方的；三次的
unbounded interval 无界区间

TEXT B

Euler's method 欧拉法
linear interpolation 线性插值
numerical estimate 数值估计
recursive scheme 递归格式
Runge-Kutta scheme 龙格-库塔格式
graphic interpretation 图解法
exact solution 精确解
vicinity 邻近；邻域
solid line 实线
piecewise-linear 分段线性
error-generating mechanism 错误生成机制
tolerance 容差
compact interval 紧区间
equidistant 等距的
refine 细化
Lipschitz function 李普希兹函数
weaker condition 弱条件
subtract 减去
Lipschitz condition 李普希兹条件
substitute 代入
first few terms 前几项
polynomial solution of degree n　n 次多项式解

slope 斜率
approximation 近似，逼近
approximant 近似式，逼近式
numerical analysis 数值分析
paradigm 范例
scalar 标量
initial condition 初始条件
dotted line 虚线
curved trajectory 曲线轨迹
numerical solution 数值解

local error 局部误差
time-stepping 时间步进的
grid 网格
step size 步长
sine qua non 〈拉〉必要条件
Taylor's theorem 泰勒定理
triangle inequality 三角不等式
inductive argument 归纳论证
replace 替换
Taylor series 泰勒级数

LESSON EIGHT

TEXT A

The Greedy Algorithm

Again, let $(E,\mathcal{F})$ be an independence system and $c: E \rightarrow \mathbb{R}_+$. We consider the MAXIMIZATION PROBLEM for $(E,\mathcal{F},c)$ and formulate two "greedy algorithms". We do not have to consider negative weights since elements with negative weight never appear in an optimum solution.

We assume that $(E,\mathcal{F})$ is given by an oracle. For the first algorithm we simply assume an **independence oracle**, i. e. an oracle which, given a set $F \subseteq E$, decides whether $F \in \mathcal{F}$ or not.

BEST-IN-GREEDY ALGORITHM
Input: An independence system $(E,\mathcal{F})$, given by an independence oracle.
　　　Weights $c: E \rightarrow \mathbb{R}_+$.
Output: A set $F \in \mathcal{F}$.

①　Sort $E = \{e_1, e_2, \cdots, e_n\}$ such that $c(e_1) \geqslant c(e_2) \geqslant \cdots \geqslant c(e_n)$.
②　Set $F: = \varnothing$.
③　**For** $i: = 1$ **to** n **do**: **If** $F \cup \{e_i\} \in \mathcal{F}$ **then** set $F: = F \cup \{e_i\}$.

The second algorithm requires a more complicated oracle. Given a set $F \subseteq E$, this oracle decides whether F contains a basis. Let us call such an oracle a **basis superset oracle**.

WORST-OUT-GREEDY ALGORITHM
Input: An independence system $(E,\mathcal{F})$ given by a basis-superset oracle.
　　　Weights $c: E \rightarrow \mathbb{R}_+$.
Output: A basis F of $(E,\mathcal{F})$

①　Sort $E = \{e_1, e_2, \cdots, e_n\}$ such that $c(e_1) \leqslant c(e_2) \leqslant \cdots \leqslant c(e_n)$.
②　Set $F: = E$.

③ **For** $i_: = 1$ **to** n **do**: If $F \setminus \{e_i\}$ contains a basis, **then** set $F_: = F \setminus \{e_i\}$.

Before we analyze these algorithms, let us take a closer look at the oracles required. It is an interesting question whether such oracles are polynomially equivalent, i. e. whether one can be simulated by a polynomial-time oracle algorithm using the other. The independence oracle and the basis-superset oracle do not seem to be polynomially equivalent:

If we consider the independence system for the TSP, it is easy to decide whether a set of edges is independent, i. e. the subset of a Hamiltonian circuit (recall that we are working with a complete graph). On the other hand, it is a difficult problem to decide whether a set of edges contains a Hamiltonian circuit.

Conversely, in the independence system for the SHORTEST PATH PROBLEM, it is easy to decide whether a set of edges contains an s-t-path. Here it is not known how to decide whether a given set is independent (i. e. subset of an s-t-path) in polynomial time.

For matroids, both oracles are polynomially equivalent. Other equivalent oracles are the **rank oracle** and **closure oracle**, which return the rank and the closure of a given subset of E, respectively.

However, even for matroids there are other natural oracles that are not polynomially equivalent. For example, the oracle deciding whether a given set is a basis is weaker than the independence oracle. The oracle which for a given $F \subseteq E$ returns the minimum cardinality of a dependent subset of F is stronger than the independence oracle.

One can analogously formulate both greedy algorithms for the MINIMIZATION PROBLEM. It is easy to see that the BEST-IN-GREEDY for the MAXIMIZATION PROBLEM for $(E, \mathcal{F}, c)$ corresponds to the WORST-OUT-GREEDY for the MINIMIZATION PROBLEM for $(E, \mathcal{F}^*, c)$: adding an element to F in the BEST-IN-GREEDY corresponds to removing an element from F in the WORST-OUT-GREEDY. Observe that KRUSKAL'S ALGORITHM is a BEST-IN-GREEDY algorithm for the MINIMIZATION PROBLEM in a cycle matroid.

The rest of this section contains some results concerning the quality of a solution found by the greedy algorithms.

Theorem 1. Let $(E, \mathcal{F})$ be an independence system. For $c: E \to \mathbb{R}_+$ we denote by $G(E, \mathcal{F}, c)$ the cost of some solution found by the BEST-IN-GREEDY for the MAXIMIZATION PROBLEM, and by $\mathrm{OPT}(E, \mathcal{F}, c)$ the cost of an optimum solution. Then

$$q(E, \mathcal{F}) \leqslant \frac{G(E, \mathcal{F}, c)}{\mathrm{OPT}(E, \mathcal{F}, c)} \leqslant 1$$

for all $c:E\rightarrow\mathbb{R}_+$. There is a cost function where the lower bound is attained.

Proof. Let $E = \{e_1,e_1,\cdots,e_n\}$, $c\colon E\rightarrow\mathbb{R}_+$, and $c(e_1)\geqslant c(e_2)\geqslant\cdots\geqslant c(e_n)$.

Let G_n be the solution found by the BEST-IN-GREEDY (when sorting E like this), while O_n is an optimum solution. We define $E_j\colon = \{e_1,\cdots,e_j\}$, $G_j\colon = G_n\cap E_j$ and $O_j\colon = O_n\cap E_j(j=0,\cdots,n)$. Set $d_n\colon = c(e_n)$ and $d_j\colon = c(e_j)-c(e_{j+1})$ for $j=1,\cdots,n-1$.

Since $O_j\in\mathcal{F}$ we have $|O_j|\leqslant r(E_j)$. Since G_j is a basis of E_j we have $|G_j|\geqslant \rho(E_j)$ With these two inequalities we conclude that

$$c(G_n) = \sum_{j=1}^{n}(|G_j|-|G_{j-1}|)c(e_j)$$

$$= \sum_{j=1}^{n}|G_j|d_j$$

$$\geqslant \sum_{j=1}^{n}\rho(E_j)d_j$$

$$\geqslant q(E,\mathcal{F})\sum_{j=1}^{n}r(E_j)d_j$$

$$\geqslant q(E,\mathcal{F})\sum_{j=1}^{n}|O_j|d_j$$

$$= q(E,\mathcal{F})\sum_{j=1}^{n}(|O_j|-|O_{j-1}|)c(e_j)$$

$$= q(E,\mathcal{F})c(O_n). \tag{1}$$

Finally we show that the lower bound is tight. Choose $F\subseteq E$ and bases B_1,B_2 of F such that

$$\frac{|B_1|}{|B_2|} = q(E,\mathcal{F}).$$

Define

$$c(e)\colon = \begin{cases} 1 & \text{for } e\in F \\ 0 & \text{for } e\in E\backslash F \end{cases}$$

and sort $e_1,\cdots,e_n$ such that $c(e_1)\geqslant c(e_2)\geqslant\cdots\geqslant c(e_n)$ and $B_1 = \{e_1,\cdots,e_{|B_1|}\}$. Then $G(E,\mathcal{F},c) = |B_1|$ and OPT$(E,\mathcal{F},c) = |B_2|$ and the lower bound is attained.

In particular we have the so-called Edmonds-Rado Theorem:

Theorem 2. *An independence system $(E,\mathcal{F})$ is a matroid if and only if the* BEST-IN-GREEDY *finds an optimum solution for the* MAXIMIZATION PROBLEM *for $(E,\mathcal{F}, c)$ for all cost functions $c:E\rightarrow\mathbb{R}_+$.*

Proof. By Theorem 1 we have $q(E,\mathcal{F})<1$ if and only if there exists a cost function $c:E\rightarrow\mathbb{R}_+$ for which the BEST-IN-GREEDY does not find an optimum solution.

We have $q(E,\mathcal{F}) < 1$ if and only if $(E,\mathcal{F})$ is not a matroid.

This is one of the rare cases where we can define a structure by its algorithmic behavior. We also obtain a polyhedral description:

Theorem 3. *Let $(E,\mathcal{F})$ be a matroid and $r:2^E \to \mathbb{Z}_+$ its rank function. Then the* **matroid polytope** *of $(E,\mathcal{F})$, i.e. the convex hull of the incidence vectors of all elements of F is equal to*

$$\left\{ x \in \mathbb{R}^E : x \geq 0,\ \sum_{e \in A} x_e \leq r(A) \text{ for all } A \subseteq E \right\}.$$

The above observation that the BEST-IN-GREEDY for the MAXIMIZATION PROBLEM for $(E, \mathcal{F}, c)$ corresponds to the WORST-OUT-GREEDY for the MINIMIZATION PROBLEM for $(E,\mathcal{F}^*,c)$ suggests the following dual counterpart of Theorem 1:

Theorem 4. *Let $(E,\mathcal{F})$ be an independent system. For $c:E \to \mathbb{R}_+$ let $G(E,\mathcal{F},c)$ denote a solution found by the WORST-OUT-GREEDY for the MINIMIZATION PROBLEM. Then*

$$1 \leq \frac{G(E,\mathcal{F},c)}{\mathrm{OPT}(E,\mathcal{F},c)} \leq \max_{F \subseteq E} \frac{|F| - \rho^*(F)}{|F| - r^*(F)} \tag{2}$$

for all $c: E \to \mathbb{R}_+$ where ρ^* and r^* are the rank functions of the dual independent system $(E,\mathcal{F}^*)$. There is a cost function where the upper bound is attained.

Proof. We use the same notation as in the proof of Theorem 1. By construction, $G_j \cup (E \backslash E_j)$ contains a basis of E, but $(G_j \cup (E \backslash E_j)) \backslash \{e\}$ does not contain a basis of E for any $e \in G_j (j = i, \cdots, n)$. In other words, $E_j \backslash G_j$ is a basis of E_j with respect to $(E, \mathcal{F}^*)$, so $|E_j| - |G_j| \geq \rho^*(E_j)$.

Since $O_n \subseteq E \backslash (E_j \backslash O_j)$ and O_n is a basis, $E_j \backslash O_j$ is independent in (E,F^*), so $|E_j| - |O_j| \leq r^*(E_j)$.

We conclude that

$$|G_j| \leq |E_j| - \rho^*(E_j) \text{ and}$$

$$|O_j| \geq |E_j| - r^*(E_j).$$

Now the same calculation as (1) provides the upper bound. To see that this bound is tight, consider

$$c(e) := \begin{cases} 1 & \text{for } e \in F \\ 0 & \text{for } e \in E \backslash F, \end{cases}$$

where $F \subseteq E$ is a set where the maximum in (2) is attained. Let B_1 be a basis of F with respect to $(E,\mathcal{F}^*)$ with $|B_1| = \rho^*(F)$. If we sort $e_1, \cdots, e_n$ such that $c(e_1) \geq c(e_2) \geq \cdots \geq$

$c(e_n)$ and $B_1 = \{e_1, \cdots, e_{|B_1|}\}$, we have $G(E, \mathcal{F}, c) = |F| - |B_1|$ and $\mathrm{OPT}(E, \mathcal{F}, c)$ $= |F| - r^*(F)$.

If we apply the WORST-OUT-GREEDY to the MAXIMIZATION PROBLEM or the BEST-IN-GREEDY to the MINIMIZATION PROBLEM, there is no positive lower/finite upper bound for $\dfrac{G(E, \mathcal{F}, c)}{\mathrm{OPT}(E, \mathcal{F}, c)}$. To see this, consider the problem of finding a minimal vertex cover of maximum weight or a maximal stable set of minimum weight in the simple graph.

However, in the case of matroids, it does not matter whether we use the BEST-IN-GREEDY or the WORST-OUT-GREEDY: since all bases have the same cardinality, the MINIMIZATION PROBLEM for $(E, \mathcal{F}, c)$ is equivalent to the MAXIMIZATION PROBLEM for $(E, \mathcal{F}, c')$, where $c'(e): = M - c(e)$ for all $e \in E$ and $M: = 1 + \max\{c(e): e \in E\}$. Therefore KRUSKAL'S ALGORITHM solves the MINIMUM SPANNING TREE PROBLEM optimally.

The Edmonds-Rado Theorem 2 also yields the following characterization of optimum k-element solutions of the MAXIMIZATION PROBLEM.

TEXT B

Computational Complexity

In contrast to the description of a problem, which is usually short, the search space is most of the time exponential in the problem dimension. In addition, for a lot of combinatorial optimization problems, one cannot hope to come up with an algorithm that produces for all problem instances an optimal solution within a time bound that is polynomial in the problem dimension. The performance measure most widely used to analyze algorithms is the time an algorithm takes to present its final answer. Time is expressed in terms of number of elementary operations such as comparisons or branching instructions. The time an algorithm needs to give the final answer is analyzed with respect to the input size. The input of a combinatorial optimization problem is often a graph or a set of integers. This input has to be represented as a sequence of symbols of a finite alphabet. The size of the input is the length of this sequence, that is, the number of symbols in it.

In this book, we are dealing with combinatorial optimization problems. Often we are considering a graph $G = (V, E)$ with n vertices and m edges and are searching for a subgraph $G' = (V', E')$ of the given one that fulfills given properties.

One approach to represent a graph is to do it by an adjacency matrix $A_G = [a_{ij}]$, where $a_{ij} = 1$ if $(v_i, v_j) \in E$ and $a_{ij} = 0$ otherwise. This matrix has n^2 entries, i. e. the number of entries is quadratic with respect to the number of vertices. An entry $a_{ij} = 1$ means that there is an edge from v_i to v_j and $a_{ij} = 0$ holds if this is not the case. Note that the adjacency matrix of a given undirected graph is symmetric. An undirected graph may have up to $\binom{n}{2} = \theta(n^2)$ edges. However, if we are considering so-called sparse graphs, the number of edges is far less than $\binom{n}{2}$.

In the case of sparse graphs, it is better to represent a given graph by so-called adjacency lists. Here, for each vertex $v \in V$ we record a set $A(v) \subseteq V$ of vertices that are adjacent to it. The size of the representation is given by the sum of the length of lists. As each edge contributes 2 to this total length, we have to write down $2m$ elements. Another factor which affects the total length of the representation is how to encode the vertices. Our alphabet has a finite size. Assume the alphabet is the set $\{0, 1\}$. Therefore we need $\theta(\log n)$ bits to encode one single vertex. This implies that we need $\theta(m \log n)$ bits (or symbols) to represent the graph G. In practice we say that a graph G can be encoded in $\theta(m)$ space, which seems to be a contradiction to the previous explanation. The reason is that computers treat all integers in their range the same. Here the same space is needed to store small integers such as 5 or large integers such as 3^{12}. We assume that graphs are considered where the number of vertices is within the integer range of the computer. This range is in most cases 0 to 2^{31}, which means that integers are represented by 32 bits. Therefore $\theta(m)$ is a reasonable approximation of the size of a graph and analyzing graph algorithms with respect to m is accepted in practice. In most cases both parameters n and m are taken into account when analyzing the complexity of a graph algorithm.

Considering graph algorithms where we can bound the runtime by a polynomial in n and m, we obviously get a polynomial-time algorithm. We have to be careful when the input includes numbers. Let $N(I)$ be the largest integer that appears in the input. An algorithm A is called pseudo-polynomial if it is polynomial in the input size $|I|$ and $N(I)$. Note that $N(I)$ can be encoded by $\theta(\log(N(I)))$ bits. Therefore a function that is polynomial in $|I|$ and $N(I)$ is not necessarily polynomial in the input size. Often the input consists of small integers. In the case where $N(I)$ is bounded by a polynomial in $|I|$, A is a polynomial-time algorithm.

An important issue that comes up when considering combinatorial optimization problems is the classification of difficult problems. To distinguish between easy and difficult problems, one considers the class of problems that are solvable by a deterministic Turing machine in polynomial time and problems that are solvable by a nondeterministic Turing machine in polynomial time. We do not want to formalize the characterization of the classes P and NP via Turing machines and prefer to outline the characteristics and notions connected with these classes at a more intuitive level. This leads to a straightforward definition to characterize problems that belong to P.

Definition 1. *A problem is in P iff it can be solved by an algorithm in polynomial time.*

Problems in P can therefore be solved in polynomial time by using an appropriate algorithm. Examples of problems belonging to this class are the single source shortest path problem and the minimum spanning tree problem.

A class that is intuitively associated with hard problems is called NP. Typically, NP is restricted to so-called decision problems, i. e. problems whose output is either YES or NO. This restriction has a technical background and captures the essentials of the problems without simplifying them too much.

Definition 2. *A decision problem is in NP iff any given solution of the problem can be verified in polynomial time.*

For problems in NP, it is therefore not necessary that a solution be computable in polynomial time. It is only necessary that we can verify the solution of the problem in polynomial time. Therefore $P \subseteq NP$ holds (slightly abusing notation by restricting P to decision problems), and it is widely assumed that $P \neq NP$.

Consider the following decision variant of the vertex cover problem. The question is whether a given graph $G = (V, E)$ contains a vertex cover of at most k vertices. Given a solution x we can easily check whether each edge is covered by x. This can be done in linear time by examining each edge at most once. Additionally, we can count the number of vertices chosen by x in linear time and therefore verify whether x is a vertex cover with at most k vertices in polynomial time.

Many optimization and decision problems, including the vertex cover problem, are at least as difficult as any problem in NP. Such problems are called NP-hard. Showing that a problem that is NP-hard is usually done by giving a polynomial-time reduction from an NP-hard problem to the considered problem. This reduction involves a transformation of the known *NP*-hard problem to the considered one, which has to be done in polynomial time. Such a reduction links the considered problem to the known *NP*-hard problem in such a way that iff the considered problem can be solved in polynomial time also the *NP*-hard problem to which it has been reduced can. We do not want to go into details and refer the reader to a book on complexity theory for further reading.

Definition 3. *A problem is called NP-hard iff it is at least as difficult as any problem in NP, i.e. each problem in NP can be reduced to it.*

As we are considering optimization problems in this book, we want to point out that many optimization problems are *NP*-hard but not in *NP*. We consider the vertex cover problem again, but at this time our main consideration is its optimization variant where the task is to compute a vertex cover of minimal size. Clearly, this optimization variant is at least as difficult as the problem of deciding whether a given graph contains a vertex cover of at most k vertices. However, since the output of the optimization problem is a number, it is not a decision problem and, therefore, not in *NP*.

In summary, many optimization problems are at least as difficult as any problem in *NP*, i.e. *NP*-hard but not in *NP*. Problems that are *NP*-hard and also in *NP* are called *NP*-complete. This holds for many decision variants of *NP*-hard optimization problems.

Definition 4. *A problem is NP-complete iff it is NP-hard and in NP.*

The classical approach to deal with *NP*-hard problems is to search for good approximation algorithms. These are algorithms that run in polynomial time but guarantee that the produced solution is within a given ratio of an optimal one. Such approximation algorithms can be totally different for different optimization problems. In the case of the *NP*-hard bin packing problem, even simple greedy heuristics work very well whereas in the case of more complicated scheduling problems methods based on linear programming are often used.

Another approach to solve *NP*-hard problems is to use sophisticated exact methods that have in the worst case an exponential runtime. The hope is that such algorithms produce good results for interesting problem instances in a small amount of time. A class of algorithms that tries to come up with exact solutions is branch and bound. Here the search space is shrunk during the optimization process by computing lower bounds on the value of an optimal solution in the case where we are considering maximization problems. The hope is to come up in a short period of time with a solution that matches such a lower bound. In this case an optimal solution has been obtained.

Related to this is the research on *parametrized complexity*. Here, parametrized versions of given optimization problems are studied. These are usually decision problems in the classical sense. Consider for example the decision variant of the vertex cover problem where we ask whether a given graph has a vertex cover of at most k vertices. This question can be answered in time $O(1.2738^k + kn)$, i. e. in polynomial time for any fixed k, and a corresponding solution with k vertices can be computed within that time bound if it exists. Obviously, this approach can be turned into an optimization algorithm that is efficient iff the value of an optimal solution is small.

A crucial consideration in combinatorial optimization problems and stochastic search algorithms that search more or less locally is the neighborhood of the current search point. Let $s \in S$ be a search point in a given search space. The neighborhood is defined by a mapping $N: S \rightarrow 2^S$. In the case we are considering combinatorial optimization problems from the search space $\{0,1\}^n$, the neighborhood can be naturally defined by all solutions having at most Hamming distance k from the current solution s. The parameter k determines the size of the neighborhood from which the next solution is sampled. Choosing a small value k, e. g. $k = 1$, such a heuristic may get stuck in local optima. If the value of k is large (in the extreme case $k = n$) and all search points of the neighborhood are chosen with the same probability, the next solution will be somehow independent of s. This leads to stochastic search algorithms that behave almost as if they were choosing in each step a search point uniformly at random from $\{0,1\}^n$. In this case the stochastic search algorithm does not take the previously sampled function values into account and the search cannot be directed into "good" regions of the considered search space.

SENTENCE STRUCTURE ANALYSIS

1. On the other hand, it is a difficult problem to decide whether a set of edges contains a Hamiltonian circuit.

另一方面，确定一组边是否包含哈密顿回路是一个困难的问题。

为了防止句子头重脚轻，这句话用 it 作形式主语，动词不定式 to decide whether a set of edges contains a Hamiltonian circuit 是真正的主语。其中，whether 引导的从句作 to decide 的宾语。

2. In contrast to the description of a problem, which is usually short, the search space is most of the time exponential in the problem dimension.

与通常简短的问题描述相反，搜索空间在问题维度中大部分都是时间指数级的。

这句话的主句是 the search space is most of the time exponential in the problem dimension。非限制性定语从句 which is usually short 修饰 the description of a problem。

注:in contrast to 与……形成对比，与……相反。

类似的表达：by contrast 相反,on the contrary 相反地,conversely 相反地

表示"与……相反/矛盾"的句子结构有：

This contradicts our assumption /the fact (that...) 这与我们的假设/事实矛盾。

…, contrary to our assumption. ……和我们的假设相反。

…, which contradicts our assumption. ……与我们的假设矛盾。

This shows a sharp/stark/striking contrast with …. 这与……形成鲜明的／明显的／显著的对比。

3. Considering graph algorithms where we can bound the runtime by a polynomial in n and m, we obviously get a polynomial-time algorithm.

鉴于在 n 和 m 中可以用一个多项式来限定运行时间的图论算法，所以很明显我们能得到一个多项式时间算法。

Considering graph algorithms where we can bound the runtime by a polynomial in n and m 是原因状语; we obviously get a polynomial-time algorithm 是主句。原因状语中 where 引导的定语从句修饰 graph algorithms。

注: 表示"鉴于，考虑到"意思的原因状语从句的表达有 seeing (that), considering (that), in view of the fact that, given (that)等。

WORDS AND PHRASES

TEXT A

greedy algorithm 贪心算法，贪婪算法 independent system 独立系统

output 输出 superset 超集

sort 排序 Hamiltonian circuit 哈密顿回路

matroid 拟阵 cycle matroid 循环拟阵

cost function 损失函数 optimum solution 最优解

lower bound 下界 rank function 秩函数

matroid polytope 拟阵多面体 convex hull 凸包

dual 对偶的

dual independence system 对偶独立系统

vertex cover 顶点覆盖

TEXT B

computational complexity 计算复杂度　　exponential 指数；指数的

combinatorial optimization problem 组合最优化问题

encode 编码　　alphabet 字母表

subgraph 子图　　sparse graph 稀疏图

pseudo-polynomial 伪多项式的；伪多项式　　deterministic 确定性的

nondeterministic 非确定性的　　characterization 特征

Turing machine 图灵机

single source shortest path problem 单源最短路径问题

minimum spanning tree problem 最小生成树问题

intuitive 直觉的　　decision variant 决策变量

vertex cover problem 顶点覆盖问题

polynomial-time reduction 多项式时间规约

transformation 变换　　complexity theory 复杂理论

approximation algorithm 近似算法　　ratio 比率，比例

bin packing problem 装箱问题，箱柜装载问题 greedy heuristic 贪婪启发式

complicated scheduling problem 复杂调度问题

linear programming 线性规划　　branch 分支

parametrized complexity 参数化复杂度　　stochastic 随机的

stochastic search algorithm 随机搜索算法　　current search point 当前搜索点

Hamming distance 汉明距离　　sampled 抽样的，采样的

APPENDIX 数学常见词汇

abbreviation 简写符号;简写

abscissa 横坐标

absolute complement 绝对补集

absolute convergent 绝对收敛

absolute error 绝对误差

absolute inequality 绝不等式

absolute maximum 绝对极大值

absolute minimum 绝对极小值

absolute monotonic 绝对单调

absolute value 绝对值

acceleration 加速度

acceleration due to gravity 重力加速度;地心
加速度

accumulative 累积的

acute angle 锐角

acute-angled triangle 锐角三角形

addition formula 加法公式

addition law 加法定律

addition law (of probability) (概率) 加法
定律

additive inverse 加法逆元;加法反元

additive property 可加性

adjacent side 邻边

adjoint matrix 伴随矩阵

algebra 代数

algebraic 代数的

algebraic cofactor 代数余子式

algebraic equation 代数方程

algebraic expression 代数式

algebraic fraction 代数分式;代数分数式

algebraic inequality 代数不等式

algebraic operation 代数运算

algorithm 算法

alternating series 交错级数

alternative hypothesis 择一假设;备择假设;
另一假设

altitude 高;高度;顶垂线;高线

ambiguous case 两义情况;二义情况

analysis 分析;解析

analytic expression 解析表达式

analytic geometry 解析几何

angle at the centre 圆心角

angle at the circumference 圆周角

angle between a line and a plane 直线与平面
的交角

angle between two planes 两平面的交角

angle bisection 角平分

angle bisector 角平分线;分角线

angle of depression 俯角

angle of elevation 仰角

angle of friction 静摩擦角;极限角

angle of greatest slope 最大斜率的角

angle of inclination 倾斜角

angle of intersection 相交角;交角

angle of projection 投射角

angle of rotation 旋转角

angle of the sector 扇形角

angle sum of a triangle 三角形内角和

anti-clockwise direction 逆时针方向;反时针
方向

anti-clockwise moment 逆时针力矩

anti-derivative 反导数

anti-logarithm 逆对数;反对数

anti-symmetric 反对称

apex 顶点

approximate value 近似值

approximation 近似;略计;逼近

Arabic system 阿拉伯数字系统

arbitrary 任意的

arbitrary constant 任意常数

arc length 弧长

arc-cosine function 反余弦函数

arc-sin function 反正弦函数

arc-tangent function 反正切函数

argument of a complex number 复数的辐角

argument of a function 函数的自变量

arithmetic mean 算术平均;等差中项

arithmetic progression 算术级数;等差级数

arithmetic sequence 等差序列

arithmetic series 等差级数

array 数组

ascending powers of X　X 的升幂

assertion 断语;断定

associative law 结合律

assumed mean 假定平均数

assumption 假定;假设

asymmetrical 非对称

asymptotic direction 渐近方向

asymptotic error constant 渐近误差常数

augmented matrix 增广矩阵

auxiliary angle 辅助角

average rate of change 平均变化率

average speed 平均速率

average value 平均值

axiomatic theory of probability 概率公理论

axis of symmetry 对称轴

band matrix 带状矩阵

bar chart 棒形图;条线图;条形图;线条图

base (1) 底;(2) 基;基数

base angle 底角

base area 底面

base line 底线

base number 底数;基数

base of logarithm 对数的底

basis 基底

Bayes' theorem 贝叶斯定理

Bernoulli distribution 伯努利分布

Bernoulli process 伯努利过程

Bernoulli trials 伯努利试验

best approximation 最佳逼近

bias 偏差;偏倚

bilinear 双线性的

bimodal distribution 双峰分布

binary system 二进制

binomial 二项式

binomial distribution 二项分布

binomial expression 二项式

binomial series 二项级数

binomial theorem 二项式定理

bisect 平分;等分

bisection method 二分法

bisector 等分线;平分线

block diagonal matrix 块对角矩阵

block matrix 分块矩阵

Boolean algebra 布尔代数

boundary condition 边界条件

boundary element 边界元素

boundary integral 边界积分

boundary line 界(线);边界

boundary point 边界点

bounded 有界的

bounded above 有上界的;上有界的

bounded below 有下界的;下有界的

bounded function 有界函数

bounded sequence 有界序列

broken line graph 折线图

Brownian motion 布朗运动

calculus of variation 变分法

Cartesian coordinates 笛卡儿坐标

Cartesian plane 笛卡儿平面

Cauchy sequence 柯西序列

Cauchy's principal value 柯西主值

Cauchy-Schwarz inequality 柯西-施瓦茨不
等式

central limit theorem 中心极限定理

centre of a circle 圆心

certain event 必然事件

chain rule 链式法则

chance 机会

change of axes 坐标轴的变换

change of base 基的变换

change of coordinates 坐标变换

change of subject 主项变换

change of variable 换元;变量替换

characteristic equation 特征方程

characteristic function 特征函数

characteristic of logarithm 对数的首数;对数
的定位部

characteristic polynomial 特征多项式

characteristic root 特征根

check digit 检验数位

chord of contact 切点弦

circulant matrix 循环矩阵

circular 圆形;圆的

circular function 圆函数;三角函数

circular measure 弧度法

circular motion 圆周运动

circular permutation 环形排列；圆形排列；
循环排列

classical theory of probability 古典概率论

clockwise direction 顺时针方向

clockwise moment 顺时针力矩

closed convex region 闭凸区域

closed interval 闭区间

closed set 闭集

closure 闭包

cluster point 聚点

coaxial 共轴

coaxial circles 共轴圆

coaxial system 共轴系

coded data 编码数据

coding method 编码法

coefficient 系数

coefficient matrix 系数矩阵

coefficient of friction 摩擦系数

coefficient of variation 变异系数

cofactor 余因子；余因式

column matrix 列矩阵

column rank 列秩

column stochastic matrix 列随机矩阵

column vector 列向量

combination 组合

combinatorial optimization 组合优化

combinatorics 组合数学

common chord 公弦

common denominator 同分母;公分母

common difference 公差

common divisor 公约数;公约

common factor 公因子;公因数

common logarithm 常用对数

common ratio 公比

complement 余;补余

complementary angle 余角

complementary equation 补充方程

complementary event 互补事件

complementary function 余函数

complementary probability 互补概率

completable 可完备化

complete oscillation 全振动

complete set 完全集

completing the square 配方

complex conjugate 复共轭

complex number 复数

complex number plane 复数平面

complex root 复数根

complexity 复杂性

component 分量

component of force 分力

composite function 复合函数；合成函数

composite number 复合数；合成数

composition of mappings 映射构合

composition of relations 复合关系

compound probability 合成概率

compound statement 复合命题；复合叙述

computational geometry 计算几何

computational science 计算科学

computer-aided design 计算机辅助设计

concave 凹的

concave downward 凹向下

concave polygon 凹多边形

concave upward 凹向上

concavity 凹性

conditional 条件句；条件式

conditional identity 条件恒等式

conditional inequality 条件不等式

condition number 条件数

conditional expectation 条件期望

conditional probability 条件概率

conditional variance 条件方差

conditionally convergent 条件收敛

cone 锥；圆锥(体)

confidence coefficient 置信系数

confidence interval 置信区间

confidence level 置信水平

confidence limit 置信极限

congruence (1) 全等；(2) 同余

congruent figures 全等图形

congruent triangles 全等三角形

conic 二次曲线；圆锥曲线

conic section 圆锥剖面

conical pendulum 圆锥摆

conjugate 共轭

conjugate axis 共轭轴

conjugate gradient method 共轭梯度法

conjugated transposed matrix 共轭转置矩阵

consequence 结论；推论

consequent 条件；后项

conservation of energy 能量守恒

conservation of momentum 动量守恒

conservative difference scheme 守恒型差分
格式

conserved 守恒

consistency condition 相容条件

consistent 一贯的；相容的

consistent estimator 相容估计量

constant 常数

constant of integration 积分常数

constraint 约束；约束条件

continuity 连续性

continuity correction 连续校正

continuous data 连续数据

continuously differentiable 连续可微的

continuous everywhere 处处连续

continuous function 连续函数

continuous proportion 连续比例

continuous random variable 连续随机变量

contradiction 矛盾

convergence in mean 平均收敛

convergence in measure 依测度收敛

convergence order 收敛阶

convergence rate analysis 收敛速度分析

convergence region 收敛域

convergent 收敛的

convergent iteration 收敛的迭代

convergent sequence 收敛序列

convergent series 收敛级数

converse 逆

converse of a relation 逆关系

converse theorem 逆定理

conversion 转换

convex analysis 凸分析

convex closure 凸闭包

convex combination 凸组合

convex cone 凸锥

convex hull 凸包

convex polygon 凸多边形

convex polyhedron 凸多面体

convex programming 凸规划

convexification 凸化

convexity 凸性

convolution 卷积

coordinate 坐标

coordinate geometry 解析几何;坐标几何

coordinate system 坐标系

coplanar 共面的

co-prime 互质；互素

corollary 推论

correct to 准确至；取值至

correlation 相关

correlation coefficient 关联系数

correspondence 对应

cosecant 余割

cosine 余弦

cosine formula 余弦公式

cotangent 余切

countable set 可数集

counter example 反例

counting 数数;计数

covariance 协方差

covariance matrix 协方差矩阵

criterion 准则

critical point 临界点

critical region 临界域

critical value 临界值

cross-multiplication 交叉相乘

cross product 向量积

cross-section 横切面;横截面;截痕

cube root 立方根

cubic 三次方;立方;三次(的)

cubic equation 三次方程

cubic roots of unity 单位的立方根

cuboid 长方体;矩体

cumulative 累积的

cumulative distribution function 累积分布函数

cumulative frequency 累积频数;累积频率

cumulative frequency curve 累积频数曲线

cumulative frequency distribution 累积频数分布

cumulative frequency polygon 累积频数多边
形;累积频率直方图

curvature of a curve 曲线的曲率

curve fitting 曲线拟合

curve sketching 曲线描绘(法)

curved line 曲线

curved surface 曲面

curved surface area 曲面面积

cylinder 柱;圆柱体

cylindric coordinate 柱坐标;柱面坐标

cylindrical 圆柱形的

D'Alembert's criterion 达朗贝尔判别法

damped oscillation 阻尼振动

Darboux upper integral 达布上积分

decagon 十边形

decay factor 衰变因子

decelerate 减速

deceleration 减速度

decile 十分位数

decimal 小数

decimal place 小数位

decimal point 小数点

decimal system 十进制

decision theory 决策论

declarative sentence 说明语句

declarative statement 说明命题

decoding 解码

decreasing function 递减函数;下降函数

decreasing sequence 递减序列;下降序列

decreasing series 递减级数;下降级数

decrement 减量

deduction 推论

deductive reasoning 演绎推理

definite 确定的;定的

definite integral 定积分

degree (1) 度;(2) 次

degree of a polynomial 多项式的次数

degree of accuracy 准确度

degree of confidence 置信度

degree of freedom 自由度

degree of ODE 常微分方程次数

degree of precision 精确度

deleted neighborhood 去心邻域

denary number 十进数

denominator 分母

dependent event(s) 相关事件;相依事件;
  从属事件

dependent variable 因变量;因变数

depreciation 折旧

derivable 可导

derivative 导数

derived curve 导函数曲线

derived function 导函数

descending order 递降序

descending powers of X X 的降序

descriptive statistics 描述统计学

detached coefficient 分离系数(法)

determinant 行列式

deviation 偏差;变差

diagonal matrix 对角矩阵

diagonalizable matrix 可对角化矩阵

diagonally dominant matrix 对角占优矩阵

difference 差

difference equation 差分方程

difference of sets 差集

difference quotient 差商

differentiable 可微的

differential 微分

differential coefficient 微商;微分系数

differential equation 微分方程

differential mean value theorem 微分中值
  定理

differentiate 求……的导数

differentiate from first principle 从基本原理
  求导数

differentiation 微分法

dimension 维数

direct proportion 正比例

directed line 有向直线

directed line segment 有向线段

directed number 有向数

direction angle 方向角

direction cosine 方向余弦

direction derivative 方向导数

direction ratio 方向比

directrix 准线

Dirichlet function 狄利克雷函数

discontinuity 不连续性

discontinuous 间断(的);连续(的);不连续(的)

discontinuous function 不连续函数

discontinuous point 间断点

discrete 分立;离散

discrete data 离散数据;间断数据

discrete process 离散过程

discrete random variable 离散随机变量

discrete uniform distribution 离散均匀分布

discretization 离散化

discriminant 判别式

disjoint 不相交的

disjoint sets 不相交的集

displacement 位移

disprove 反证

distance 距离

distance formula 距离公式

distinct roots 相异根

distinct solution 相异解

distributive law 分配律

diverge 发散

divergence 发散(性)

divergent 发散的

divergent iteration 发散性迭代

divergent sequence 发散序列

divergent series 发散级数

divide 除

dividend (1) 被除数;(2) 股息

divisible 可整除

division 除法

division algorithm 除法算式

divisor of zero 零因子

dodecagon 十二边形

domain decomposition method 区域分解法

double angle formula 二倍角公式

double root 二重根

double integral 二重积分

doubly stochastic matrix 双随机矩阵

dual optimal solution 对偶最优解

duality 对偶性

dynamic optimization 动态最优化

dynamic programming 动态规划

eccentric angle 离心角

echelon matrix 梯矩阵

edge 棱;边

efficient estimator 有效估计量

eigenspace 特征空间

eigenvalue 特征值

eigenvector 特征向量

elastic body 弹性体

elementary event 基本事件

elementary function 初等函数

elementary row operation 基本行运算

elimination 消法

elimination method 消去法;消元法

ellipse 椭圆

ellipsoid 椭圆体

ellipsoidal algorithm 椭球算法

elliptic function 椭圆函数

elongation 伸张;展

empirical data 实验数据

empirical formula 实验公式

empirical probability 实验概率;经验概率

empty set 空集

encoding 编码

enclosure 界限

end point 端点

entire surd 整方根

equal ratios theorem 等比定理

equality constraint 等式约束

equality sign 等号

equation in one unknown 一元方程

equation in two unknowns（variables）二元方程

equation of a straight line 直线方程

equilateral polygon 等边多边形

equilateral triangle 等边三角形

equilibrium point 平衡点

equiprobable 等概率的

equiprobable space 等概率空间

equivalence 等价

equivalence class 等价类

equivalence relation 等价关系

equivalent 等价（的）

error 误差

error allowance 容许误差

error estimate 误差估计

error term 误差项

error tolerance 误差宽容度

escribed circle 旁切圆

estimate 估计；估价

estimator 估计量

Euclidean algorithm 欧几里得算法

Euclidean geometry 欧几里得几何

Euler's formula 欧拉公式

evaluate 计值

even function 偶函数

even number 偶数

evenly distributed 均匀分布的

event 事件

exact differential form 恰当微分形式

exact solution 准确解；精确解；真确解

exact value 准确值；精确值；真确值

exclusive events 互斥事件

expand form 展开式

expectation value/expected value 期望值；预期值

explicit function 显函数

exponent 指数

exponential function 指数函数

exponential order 指数阶；指数级

express…in terms of… 以……表达

expression 式；数式

extension 外延；延长；扩张；扩充

extension of a function 函数的扩张

extreme point 极值点

extreme value 极值

extremum 极值

face 面

factor 因子；因式；商

factor theorem 因子定理；因式定理

factorial 阶乘

factorization 因子分解；因式分解

factorization method 因式分解法

factorization of polynomial 多项式因式分解

family of straight lines 直线族

feasible set 可行集

feasible solution 可行解；容许解

Fermat last theorem 费马大定理

Fibonacci number 斐波那契数；黄金分割数

Fibonacci sequence 斐波那契序列

figure （1）图（形）；（2）数字

final velocity 末速度

finite difference 有限差分

finite dimensional vector space 有限维向量空间

finite probability space 有限概率空间

finite sequence 有限序列

finite series 有限级数

finite set 有限集

first derivative 一阶导数

first order differential equation 一阶微分方程

first projection 第一投影；第一射影

first quartile 第一四分位数

first term 首项

fixed point 定点

fixed point iteration method 定点迭代法

for all X 对所有 X

for each /every X 对每一 X

formula(formulae) 公式

Fourier series 傅立叶级数

Fourier transform 傅立叶变换

fourth root 四次方根

fraction 分数；分式

fraction in lowest term 最简分数

fractional equation 分式方程

fractional index 分数指数

fractional inequality 分式不等式

frequency distribution 频数分布；频率分布

function 函数

function of bounded variation 有界变差函数

function of function 复合函数；迭函数

functional notation 函数记号

function series 函数级数

fundamental solution 基本解

fundamental theorem of algebra 代数基本定理

fundamental theorem of calculus 微积分基本定理

Gaussian distribution 高斯分布

Gaussian elimination 高斯消去法

Gauss quadrature 高斯求积

general form 一般式；通式

general solution 通解；一般解

generalized eigenspace 广义本征空间

generalized eigenvector 广义本征向量

generalized integral 广义积分

generalized inverse matrix 广义逆矩阵

generating function 母函数；生成函数

geometric distribution 几何分布

geometric programming 几何规划

geometric mean 几何平均数；等比中项

geometric progression 等比数列

geometric sequence 等比序列

geometric series 等比级数

global extreme value 整体极值

global maximum 全局极大值；整体极大值

global minimum 全局极小值；整体极小值

golden section 黄金分割

gradient (1) 斜率；倾斜率;(2) 梯度

gradient vector 梯度向量

grand total 总计

graphical solution 图解

greatest lower bound 最大下界

Green formula 格林公式

grid lines 网格线

grouped data 分组数据；分类数据

grouping terms 并项；集项

growth factor 增长因子

half closed interval 半闭区间

half open interval 半开区间

harmonic mean (1) 调和平均数；(2) 调和中项

harmonic series 调和级数

harmonics 调和函数

heuristic analysis 启发式分析

higher order derivative 高阶导数

highest common factor(H. C. F) 最大公因子；最高公因式；最高公因子

Hindu-Arabic numeral 阿拉伯数字

histogram 组织图；直方图；矩形图

Hölder's Inequality 赫尔德不等式

homogeneous 齐次的

homogeneous equation 齐次方程

Hooke's law 胡克定律

horizontal asymptote 水平渐近线

horizontal component 水平分量

horizontal line 横线;水平线

horizontal range 水平距离,范围

hyperbola 双曲线

hyperbolic function 双曲函数

hypergeometric distribution 超几何分布

hyperplane 超平面

hypothesis 假设

hypothesis testing 假设检验

hypothetical syllogism 假设三段论

idempotent matrix 幂等矩阵

identical 全等;恒等

identity 等(式)

identity element 单位元

identity law 同一律

identity mapping 恒等映射

identity matrix 单位矩阵

identity relation 恒等关系式

if and only if/iff 当且仅当;若且仅若

if…, then 若……则;如果……则

ill-conditioned problem 病态问题

illustration 例证;说明

image 像点;像

imaginary number 虚数

imaginary part 虚部

imaginary root 虚根

imaginary unit 虚数单位

implicit differentiation 隐函数微分

implicit function 隐函数

improper integral 反常积分

inclined plane 斜面

included angle 夹角

included side 夹边

inclusion mapping 包含映射

inclusive 包含的;可兼的

inclusive disjunction 包含性析取;可兼析取

inconsistent 不一致(的)

increasing function 递增函数

increasing sequence 递增序列

increasing series 递增级数

increment 增量

indecomposable matrix 不可分解矩阵

indefinite integral 不定积分

indefinite integration 不定积分法

indefinite matrix 不定矩阵

independence 独立;自变

independent equations 独立方程

independent random variable 独立随机变量

independent variable 自变量;独立变量

indeterminate (1) 不定的;(2) 不定元;未定元

indeterminate coefficient 不定系数;未定系数

indeterminate form 待定型;不定型

index 指数

induced set 导集

induction hypothesis 归纳法假设

inelastic collision 非弹性碰撞

inequality constraint 不等式约束

inequality sign 不等号

inertia 惯性;惯量

inference 推论

infimum 下确界

infinite dimensional 无穷维的

infinite population 无限总体

infinite sequence 无限序列;无穷序列

infinite series 无限级数;无穷级数

infinitely many 无穷多

infinitesimal 无限小;无穷小

infinity 无限(大);无穷(大)

inflection point 拐点;转折点

inherent error 固有误差

inhomogeneous equation 非齐次方程

initial approximation 初始近似值

initial condition 原始条件;初值条件

initial point 始点;起点

initial side 始边

initial velocity 初速度

initial-value problem 初值问题

injective function 内射函数

inner product 内积

insertion 插入

insertion of brackets 加括号

instantaneous 瞬时的

instantaneous acceleration 瞬时加速度

instantaneous rate of change 瞬时变化率

instantaneous speed 瞬时速率

instantaneous velocity 瞬时速度

integer programming 整数规划

integrability condition 可积性条件

integrable 可积的

integrable function 可积函数

integral geometry 积分几何

integral index 积分指数

integral mean value theorem 积数中值定理

integral part 整数部分

integral sign 积分号

integral solution 整数解

integral test 积分检验法

integral value 整数值

integrand 被积函数

integrate 积;积分;……的积分

integrating factor 积分因子

integration 积分法

integration by parts 分部积分法

integration by substitution 代换积分法;换元
  积分法

integration constant 积分常数

interaction 相互作用

intermediate value theorem 介值定理

internal bisector 内分角线

internal division 内分

internal point of division 内分点

interpolating polynomial 插值多项式

interpolation 插值

inter-quartile range 四分位数间距

intersect 相交

intersection (1) 交集;(2) 相交;(3) 交点

interval 区间

interval estimation 区间估计;区域估计

invariance 不变性

invariant linear subspace 不变线性子空间

inverse circular function 反三角函数

inverse cosine function 反余弦函数

inverse function 反函数;逆函数

inverse mapping 反向映射;逆映射

inverse matrix 逆矩阵

inverse problem 逆算问题

inverse proportion 反比;逆比例

inverse relation 逆关系

inverse sine function 反正弦函数

inverse tangent function 反正切函数

inverse variation 反变(分);逆变(分)

invertible 可逆的

invertible matrix 可逆矩阵

irrational equation 无理方程

irrational number 无理数

irreducibility 不可约性

irreducible matrix 不可约矩阵

isolated point 孤立点

isomorphism 同构

isosceles triangle 等腰三角形

isotone 保序

iterate（1）迭代值；（2）迭代

iterated integral 逐次积分

iteration 迭代

iteration form 迭代形

iterative function 迭代函数

iterative method 迭代法

Jacobian matrix 雅可比矩阵

joint distribution function 联合分布函数

joint variation 联变（分）；连变（分）

jump discontinuity 跳跃间断性

L'Hospital's rule 洛必达法则

Lagrange interpolating polynomial 拉格朗日插值多项代

Lagrange multiplier 拉格朗日乘子

Lagrange theorem 拉格朗日定理

Laplace expansion 拉普拉斯展开式

law of indices 指数律；指数定律

law of inference 推论律

leading coefficient 首项系数

least common multiple 最小公倍数

least squares method 最小二乘法

least squares problem 最小二乘问题

least value 最小值

left continuous 左连续

left derivative 左导数

left hand limit 左方极限

lemma 引理

lemniscate 双纽线

limit of sequence 序列的极限

limiting case 极限情况

limiting position 极限位置

line integral 线积分

line of best-fit 最佳拟合

line of greatest slope 最大斜率线

line of intersection 交线

line segment 线段

linear 线性；一次

linear combination 线性组合

linear complementarity 线性互补

linear convergence 线性收敛

linear dependence 线性相关

linear differential equation 线性微分方程

linear equation 线性方程；一次方程

linear equation in two unknowns 二元一次方程；二元线性方程

linear independence 线性无关

linear inequality 一次不等式；线性不等式

linear programming 线性规划

linear regression 线性回归

linear subspace 线性子空间

linear transformation 线性变换

linearly dependent 线性相关的

linearly independent 线性无关的

local extreme value 局部极值

local maximum 局部极大（值）

local minimum 局部极小（值）

locus 轨迹

logarithm 对数

logarithmic differentiation 对数求导法

logarithmic equation 对数方程

logarithmic function 对数函数

logical deduction 逻辑推论；逻辑推理

lower bound 下界

lower limit 下限

lower triangular matrix 下三角形矩阵

lowest common multiple（L. C. M）最小公倍数

Maclaurin expansion 麦克劳林展开式

Maclaurin series 麦克劳林级数

magnitude 量；数量；长度；大小

main diagonal 主对角线

mantissa of logarithm 对数的尾数；对数的定

值部

many to one 多对一

many-sided figure 多边形

many-valued 多值的

mapping 映射

Markov chain 马尔可夫链

martingale 鞅

mathematical analysis 数学分析

mathematical induction 数学归纳法

mathematical programming 数学规划

matrix 阵；矩阵

matrix addition 矩阵加法

matrix equation 矩阵方程

matrix multiplication 矩阵乘法

matrix operation 矩阵运算

maximal eigenvalue 最大特征值

maximal value 极大值

maximize 极大

maximum absolute error 最大绝对误差

maximum likelihood 最大似然

maximum point 极大点

mean 平均(值)；平均数；中数

mean deviation 平均偏差

mean value theorem 中值定理

measurement 量度

mensuration 计量；求积法

method of completing square 配方法

method of interpolation 插值法；内插法

method of least squares 最小二乘法；最小平
方法

method of substitution 代换法；换元法

method of successive substitution 逐次代换
法；逐次调替法

method of superposition 叠加法

metric unit 十进制单位

mid-point 中点

mid-point formula 中点公式

mid-point theorem 中点定理

minimal norm solution 极小范数解

minimal polynomial 最小多项式

minimax theorem 极小极大定理

minimum point 极小点

minimum value 极小值

Minkowski Inequality 闵可夫斯基不等式

minor (1)子式；(2)较小的

minor of a determinant 子行列式

model 模型

modular (1)模；模数；(2)同余

modulo arithmetic 同余算术

modulus 模量；模；模数

modulus of a complex number 复数的模

monomial 单项式

monotone 单调

monotone function 单调函数

monotonic convergence 单调收敛性

monotonic decreasing 单调递减

monotonic decreasing function 单调递减函数

monotonic increasing 单调递增

monotonic increasing function 单调递增函数

Monte Carlo method 蒙特卡罗方法

multinomial 多项式

multiple 倍数

multiple angle 倍角

multiple-angle formula 倍角公式

multiple coefficient of correlation 多重相关
系数

multiple correlation 多重相关

multiple integral 多重积分

multiple root 多重根

multiplicand 被乘数

multiplication 乘法

multiplication law (of probability) (概率)乘

法定律

multiplicative property 可乘性

multiplicity 重数

multiplier 乘数；乘式

multi-value 多值的

multivariable function 多元函数

multivariate statistical analysis 多元统计分析

mutually disjoint 互不相交

mutually exclusive events 互斥事件

mutually independent 独立；互相独立

mutually perpendicular lines 互相垂直

n factorial n 阶乘

nth derivative n 阶导数

nth root n 次根；n 次方根

n the root of unity 单位的 n 次根

Napierian logarithm 纳皮尔对数；自然对数

natural exponential function 自然指数函数

natural logarithm 自然对数

natural number 自然数

natural surjection 自然满射

necessary and sufficient condition 充要条件

necessary condition 必要条件

negation 否定式

negative angle 负角

negative binomial distribution 负二项式分布

negative definite matrix 负定矩阵

negative index 负指数

negative integer 负整数

negative number 负数

negative vector 负向量；负矢量

neighborhood 邻域

Newton's formula 牛顿公式

Newton's method 牛顿方法

nilpotent 幂零的

nondifferentiable point 不可微点

non-linear 非线性

non-linear equation 非线性方程

nonnegative matrix 非负矩阵

non-reflexive 非自反的

nonsingular matrix 非奇异矩阵

nonsmooth analysis 非光滑分析

non-transitive 非可递的

non-trivial 非平凡的

non-zero 非零

norm 范数

normal (1) 垂直的；正交的；法线的；(2) 正态的；(3) 正常的；正规的

normal curve 标准曲线

normal distribution 正态分布

normal matrix 正规矩阵

normal vector 法向量

normalize 正规化

normalized form 标准型

notation 记法；记号

nowhere dense 稀疏的

null 零；空

null hypothesis 零假设；虚假设

null set 空集

null vector 零向量

numerical analysis 数值分析

numerical differentiation 数值微分

numerical expression 数字式

numerical integration 数值积分

numerical scheme 数值格式

objective function 目标函数

oblique asymptote 斜渐近线

obtuse angle 钝角

obtuse-angled triangle 钝角三角形

odd function 奇函数

odd number 奇数

one-sided limit 单侧极限

one-to-many 一对多

one-to-one 一对一

one-one correspondence 一一对应

one-one mapping 满射

one-variable function 一元函数

open(closed) ball 开(闭)球

open domain 开域

open interval 开区间

operations research 运筹学

opposite angle 对角

optimal solution 最优解

optimality condition 最优性条件

order (1) 序;次序;(2) 阶;级

order of a matrix 矩阵的阶

ordered n-tuples 有序 n 元;有序 n 配列

ordered pair 序偶

ordered relation 有序关系

ordered set 有序集

ordinal number 序数

ordinary differential equation 常微分方程

ordinate 纵坐标

origin 原点

orthocentre 垂心

orthogonal basis 正交基

orthogonal circles 正交圆

orthogonal complement 正交补

orthogonal decomposition 正交分解

orthogonal matrix 正交矩阵

orthogonal projection 正射影

orthogonal transformation 正交变换

orthogonality 正交性

parabola 抛物线

paraboloid 抛物面

paradox 悖论

parallel 平行(的)

parameter 参数;参变量

parametric equation 参数方程

parametric form 参数式

partial derivative 偏导数

partial differential 偏微分

partial fraction 部分分式

partial sum 部分和

partial variation 部分变(分)

particular solution 特解

partition of unity 单位分解

pattern 模型;规律

percentage decrease 百分减少

percentage error 百分率误差

percentage increase 百分增加

percentile 百分位数

perfect set 完全集

perfect square 完全平方

perimeter 周长;周界

periodic function 周期函数

permutation 排列

permutation with repetition 重复排列

permutation without repetition 无重复排列

perpendicular 垂线;垂直(于)

perpendicular bisector 垂直平分线;中垂线

perpendicular line 垂直线

phase 相;位相

phase shift 相移

piecewise linear function 分段线性函数

pivot 支点

plane 平面

plane figure 平面图形

plot 绘图

point circle 点圆

point estimation 点估计

point of contact 切点

point of division 分点

point of inflection (inflexion) 拐点;转折点

point of intersection 交点

point-slope form 点斜式

Poisson distribution 泊松分布

polar axis 极轴

polar coordinate 极坐标系

polar coordinates 极坐标

polar coordinate plane 极坐标平面

polar coordinate system 极坐标系统

polar equation 极方程

polar form 极形式

polygon 多边形

polygon law of addition 多边形加法

polygon method 多边形法

polyhedron 多面体

polynomial 多项式

polynomial equation 多项式方程

population 总体

population mean 总体平均(值)

position vector 位置向量；位置矢量

positive definite 正定的

positive index 正指数

positive integer 正整数

positive matrix 正矩阵

positive number 正数

positive semidefinite matrix 正半定矩阵

posterior probability 后验概率；事后概率

post-multiply 后乘；自右乘

postulate 公理

potential energy 势能；位能

potential function 势函数

power (1) 幂；乘方；(2) 功率；(3) 检定力

power function 幂函数

power series 幂级数

power set 幂集；势集

precision 精确度

prime factor 质因子；质因素

prime number 素数；质数

primitive (1) 本原的；原始的；(2) 原函数

primitive function 原函数

primitive matrix 本原矩阵；素矩阵

primitive quadratic form 本原二次型

principal value 主值

principal value interval 主值区间

prior probability 先验概率；事先概率

prism 棱柱(体)；角柱(体)

prismoid 平截头棱锥体

probability density function 概率密度函数

probability distribution 概率分布

probability generating function 概率母函数

probability method 概率方法

probability theory 概率论

progression 级数

projecting plane 投射平面

projection matrix 投影矩阵

proof by contrapositive 反证法

proper fraction 真分数

proper integral 正常积分

proper subset 真子集

property 性质

proportional 成比例

proposition 命题

propositional calculus 命题演算

propositional inference 命题推演

pseudoconvex function 伪凸函数

purely imaginary number 纯虚数

Pythagoras theorem 勾股定理

quadrant 象限

quadratic convergence 二阶收敛性

quadratic equation 二次方程(式)

quadratic form 二次型

quadratic formula 二次公式

quadratic function 二次函数

quadratic inequality 二次不等式

quadratic polynomial 二次多项式

quadratic programming 二次规划

quadrature 求积法

quasiconvex function 拟凸函数

quasi Newton method 拟牛顿法

queuing theory 排队论

radian measure 弧度法

radical 根式;根号;根数

random distribution 随机分布

random event 随机事件

random experiment 随机试验

random number 随机数

random number generator 随机数产生器

random sample 随机样本

random variable 随机变量

random walk 随机游走

randomness 随机性

range 值域;区域;范围;极差;分布域

rank 秩

rate of change 变化率

rate of convergence 收敛率

rational expression 有理式;有理数式

rational function 有理函数

rational index 有理数指数

rational number 有理数

rationalization 有理化

real axis 实轴

real number 实数

real part 实部

real root 实根

rearrangement of series 级数的重排

rectangular coordinate plane 直角坐标平面

rectangular coordinates 直角坐标

rectangular distribution 矩形分布

rectangular formula 矩形公式

rectangular matrix 长方阵

rectangular number 矩形数

rectifiable 可求长的

recurrence formula 递推公式

recurrent 循环的

recurring decimal 循环小数

reducible 可约的;可化简的

reflexive relation 自反关系

region of acceptance 接受区域

region of convergency 收敛区域

region of rejection 拒绝区域

regular polygon 正多边形

reject 舍去;否定

relation 关系;关系式

relative error 相对误差

relative frequency 相对频数

relative interior point 相对内点

relative maximum 相对极大

relative minimum 相对极小

relative motion 相对运动

relative velocity 相对速度

relatively prime 互素

remainder 余数;余式;剩余

remainder term 余项

remainder theorem 余式定理

removal of brackets 撤括号;去括号

repeated integral 累次积分

repeated trials 重复试验

residual 残差;剩余

resolution of vector 向量分解;矢量分解

resolve 分解

rhombus 菱形

Riemann integral 黎曼积分

right angle 直角

right circular cone 直立圆锥(体)

right circular cylinder 直立圆柱(体)

right hand limit 右方极限

right-angled triangle 直角三角形

Rolle's theorem 罗尔定理

root 根

root-mean-square 均方根

rounding off 舍入；四舍五入

round-off error 舍入误差

row 行

row rank 行秩

row vector 行向量；行矢量

sample 抽样；样本

sample mean 样本均值

sample space 样本空间

sampling distribution 抽样分布

sampling theory 抽样理论

sandwich theorem 迫近定理

scalar 标量

Schwartz's inequality 施瓦茨不等式

scientific notation 科学记数法

secant（1）正割；（2）割线

secant method 正割法

second derivative 二阶导数

second order ordinary differential equation 二
　阶常微分方程

second quartile 第二四分位数

section（1）截面；截线；（2）截点

section formula 截点公式

sector 扇式

segment 段；节

semi-conjugate axis 半共轭轴

semidefinite programming 半定规划

semi-major axis 半主轴；半长轴

semi-martingale 半鞅

separable differential equation 可分微分方程

sequence 序列

series 级数

series of functions 函数级数

series of positive terms 正项级数

set 集

set square 三角尺；三角板

set theory 集合论

several variables 多变量

signed number 有符号数

significant figure 有效数字

signum 正负号函数

simple iteration method 简单迭代法

simulation 模拟

simultaneous approximation 同时逼近

simultaneous differential equations 微分方程
　组；联立微分方程

simultaneous equations 联立方程

simultaneous inequalities 联立不等式

simultaneous linear equations in two unknowns
　联立二次线性方程组

sine 正弦

sine formula 正弦公式

singleton 单元集

single-valued function 单值函数

singular 奇数的

singular matrix 奇异矩阵；不可逆矩阵

skew distribution 偏斜分布

skew line 偏斜线

slack variable 松弛变量

smooth function 光滑函数

solution 解；解法

solution of equation 方程解

solution of triangle 三角形解法

solution set 解集

solve 解

space curve 空间曲线

spanned by... 由……生成

sparse matrix 稀疏矩阵

special angle 特殊角；特别角

spectral method 谱方法

spectral radius 谱半径

spectrum 谱

spherical coordinates 球面坐标

spheroid 球体

spiral 螺线

spline function 样条函数

square bracket 方括号

square matrix 方阵

square number 平方数

square root 平方根；二次根

squeeze theorem 迫近定理

stability 稳定性

standard deviation 标准差；标准偏离

standard equation 标准方程

standard error 标准误差

standard form 标准式

standard normal distribution 标准正态分布；
  标准常态分布

standard score 标准分

standard unit 标准单位

statement 语句

statement calculus 命题演算

stationary point 平稳点；逗留点；驻点

stationary value 平稳值

statistical chart 统计图

statistical data 统计数据

statistical error 统计误差

statistical hypothesis 统计假设

statistical significance 统计显著性

statistics 统计

steepest descent 最速下降

step function 阶梯函数

stochastic convergence 随机收敛

stochastic differential 随机微分

stochastic process 随机过程

stochastic programming 随机规划

stochastic variable 随机变量

stochastic sampling 随机取样

Stokes' theorem 斯托克斯定理

straight line graph 直线图像

strictly convex function 严格凸函数

strictly monotonic 严格单调

strictly monotonic function 严格单调函数

structure preserving method 保结构方法

subject 主项

submultiple angle formula 半角公式

subnormal 次法线

subsequence 子(序)列

subscript 下标

subgradient method 次梯度方法

subset 子集

subsidiary angle 辅助角

substitute 代入

substitution 代入；代入法

subtangent 次切线

subtraction 减法

successive approximation 逐次逼近法

successive derivative 逐次导数

successive differentiation 逐次微分法

sufficiency 充分性

sufficient and necessary condition 充要条件

sufficient condition 充分条件

sufficiently close 充分接近

sum to infinity 无限项之和

sum to n terms n 项和

sum to product formula 和化积公式

summation 求和法；总和

summation formula 求和公式

superimposing 迭加

superscript 上标

super set 超集

supremum value 上确界

supplementary angle 补角

surface area 表面面积；曲面面积

surface integral 面积分

symmetric matrix 对称矩阵

symmetric relation 对称关系

symmetry 对称；对称性

synthetic division 综合除法

symplectic algorithm 辛算法

tabulation form 表列式

tabulation method 列表法

tangent （1）正切；（2）切线

tautology 恒真命题；恒真式

Taylor's expansion 泰勒展开式

Taylor's series 泰勒级数

Taylor's theorem 泰勒定理

tension 张力

term 项

terminal point 终点

terminal side 终边

test criterion 检验标准

test of significance 显著性检验

theorem 定理

theoretical probability 理论概率

three-dimensional space 三维空间

total differential 全微分

total probability 总概率

touch 切；切触

trajectory 轨；轨迹

transcendental function 超越函数

transcendental number 超越数

transitive property 传递性

transitivity 传递性

translation 平移

transpose 移项；转置

transpose of matrix 倒置矩阵；转置矩阵

transposed matrix 转置矩阵

trapezium 梯形

trapezoidal integral 梯形积分

trapezoidal rule 梯形法则

triangle inequality 三角不等式

triangle law of addition 三角形加法

triangle method 三角形法

triangular matrix 三角矩阵

triangular number 三角形数

trigonometric function 三角函数

trigonometric identity 三角恒等式

triple integral 三重积分

triple product 三重积

trisect 三等分

trust region method 信赖域法

two-dimensional space 二维空间

unbiased estimator 无偏估计量

unbounded function 无界函数

uncountable set 不可数集

undefined 未下定义（的）

undetermined coefficient 待定系数

ungrouped data 未分组数据

uniform 一致（的）；均匀（的）

uniform acceleration 匀加速度

uniform body 均匀物体

uniform cross-section 均匀横切面

uniform distribution 均匀分布

uniform motion 匀速运动

uniform speed 匀速率

uniform velocity 匀速度

uniformly bounded 一致有界的

uniformly continuous 一致连续的

uniformly convergent 一致收敛的

uniformly distributed 均匀分布的

unimodal distribution 单峰分布

union 并集

unique solution 唯一解

uniqueness 唯一性

unit area 单位面积

unit circle 单位圆

unit imaginary number 单位虚数

unit matrix 单位矩阵

unit vector 单位向量；单位矢量

unit volume 单位体积

unitary matrix 酉矩阵

universal quantifier 全称量词

upper bound 上界

upper(lower) limit 上(下)极限

upper triangular matrix 上三角形矩阵

validity 真实性；有效性

variability 可变性；变异性

variable 变项；变量；元；变元；变数

variable speed 可变速率

variable velocity 可变速度

variance 方差

variational inequality 变分不等式

vector 向量；矢量

vector addition 向量和；矢量和

vector equation 向量方程；矢量方程

vector function 向量函数；矢量函数

vector product 矢量积；矢量积

vector space 向量空间

vector subspace 向量子空间

vector triple product 向量三重积

velocity 速度

verify 证明；验证

vertex，vertices 顶(点)；极点

vertical 铅垂；垂直

vertical angle 顶角

vertical asymptote 垂直渐近线

vertical component 垂直分量

vertical line 纵线；铅垂

vertically opposite angles 垂直对顶角

void set 空集

volume of revolution 旋转体的体积

wavelet theory 小波理论

weighted average/weighted mean 加权平均数

white noise 白噪声

Wiener process 维纳过程

without loss of generality 不失一般性

x-axis x 轴

x-coordinate x 坐标

x-intercept x 轴截距

y-axis y 轴

y-coordinate y 坐标

y-intercept y 轴截距

zero element 零元素

zero factor 零因子

zero matrix 零矩阵

zero vector 零向量

zeroes of a function 函数零值

REFERENCES

［1］齐玉霞. 英汉数学词汇. 2 版. 北京:科学出版社,1982.

［2］汤涛,丁玖. 数学之英文写作. 北京:高等教育出版社,2013.

［3］杨自辰,杨大成. 科技英语写作. 北京:国防工业出版社,1991.

［4］Kostrikin A I, Manin Y I. Linear algebra and geometry. Gordon and Breach Science Publishers, 1997.

［5］Duistermatt J J, Kolk J A C. Multidimensional real analysis, Ⅱ. Cambridge: Cambridge University Press, 2004.

［6］M. Marcus. Finite Dimensional Multilinear Algebra, Part Ⅱ. New York: Marcel Dekker, 1973.

［7］Axler S. Linear algebra done right. New York: Springer, 1997.

［8］Horn R A, Johnson C R. Matrix analysis. 2th ed. Cambridge: Cambridge University Press, 2013.

［9］Lay D C. Linear algebra and its applications. 3th ed. ［S. l.］: Addison-Wesley, 2003.

［10］Axelson O. Iterative Solution methods. Cambridge: Cambridge University Press. 1996.

［11］Zeidler E. Nonlinear functional analysis and its applications（Ⅱ）: Linear monotone operators. Heidelberg: Springer,1990.

［12］Garling J H D. A course in mathematical analysis（Volume 1）: Foundations and elementary real analysis. Cambridge:Cambridge University Press, 2013.

［13］Iserles A. A first course in the numerical analysis of differential equations. Cambridge:Cambridge University Press, 2009.

［14］Simmons G B, Krantz S G. Differential Equations: theory, technique, and practice. New York: McGraw-Hill, 2006.

［15］Grafakos L. Classical fourier analysis. 3 版. 北京:世界图书出版公司, 2014.

［16］Apostol T M. 数学分析(原书:第 2 版). 邢富中,邢辰,李松洁,等译. 北京: 机械工业出版社, 2006.

［17］Neumann F, Witt C. Combinatorial optimization and computational complexity. Berlin: Springer Berlin Heidelberg, 2010.

［18］Borwein J M, Lewis A S. Convex analysis and nonlinear optimization. ［S.l.］: Springer, 2000.

［19］Roger A, Horn R A, Johnson C R. 矩阵分析. 英文版. 北京:人民邮电出版社, 2005.